A Guide to Maps & Models for UX Designers

A mini guide on mapping & modeling to understand and communicate complexity

by erin malone, mfa

A Guide to Maps & Models for UX Designers
A mini guide on mapping & modeling to
understand and communicate complexity

by erin malone, mfa

An 8 Paw Press publication
© 2023
http://8pawpress/books

ISBN: 979-8-9868133-6-3

CONTENTS

Maps and models help us see and visualize all sorts of complex concepts. They help us to get our hands and heads around a topic or a complex system, visualize an interaction, site structure, or a user's experience as well as work through ideas and simple systems.

We can model conceptual ideas, real ecosystems, objects and interactions. To help understand the boundaries of a topic or the landscape of a system we make models to visualize the unseen and scope their edges.

When you just start out, it's good to make diagrams and maps that explore those boundaries and to capture all the potential elements, actors, agents, and components, found within a system or a topic.

This mini guide is intended to be an overview of the landscape of different kinds of maps and models. It is not exhaustive and I recommend Jim Kalbach's book *Mapping Experiences* and Abby Covert's book *Stuck* for more details on many of these types of diagrams as well as for details on the missing types.

I originally gathered together all these different kinds of diagrams for my students to help them differentiate between them as I was teaching and to help them understand when you might use each kind.

The first three types of maps in this guide can help you get started when trying to get your arms around what a topic or system boundary should be.

The mind map, the cluster map and the concept map are three great starter maps and each informs the next to help you understand a topic or system.

By making these diagrams we can determine what is in or out, what elements or actors we need to pay attention to and we can start to see relationships between the concepts of a topic or system.

The later models show how you might visualize the taxonomy in a system and how to map user and system tasks, and experience journeys. The radial time and task diagrams and service blueprints explore what people actually need to do at different stages in a life cycle.

Other diagrams include sitemaps, timelines, schedules, data (information) models and other kinds of information that when presented visually can become more clear and help you have tactical discussions with development and content teams.

MIND MAPS

WHAT IS IT?
A mind map is a free-form mapping of concepts that start from a single concept in the middle and radiate out as the idea triggers new thoughts, related concepts and other ideas. The mind map is good for idea generation, poking at the edges of a main theme from different perspectives and depths and for seeing connections and patterns between concepts that might not otherwise appear related.

WHEN DO YOU DO ONE?
Early in the process to explore possibilities through non-linear thinking and to generate ideas and explore everything within a problem space.

Start with the mind map to capture everything you can think of related to the topic or idea, then move to a cluster map to help organize the ideas into meaningful groupings and start fleshing out relationships and then to do a concept map to help further understand the relationships between the ideas, actors and elements, and to refine the important threads in the system or topic.

Many people have experience making mind maps and this is one of the easiest to get your head around. We do them when we brainstorm and its a good way to try and pull out of our heads and our research absolutely everything we can think of about a topic. It also allows us to free associate and to rabbit hole without negative consequences. Mind maps are good thinking tools for ourselves and our teams. It's important to start quickly and be free with your thinking without censorship.

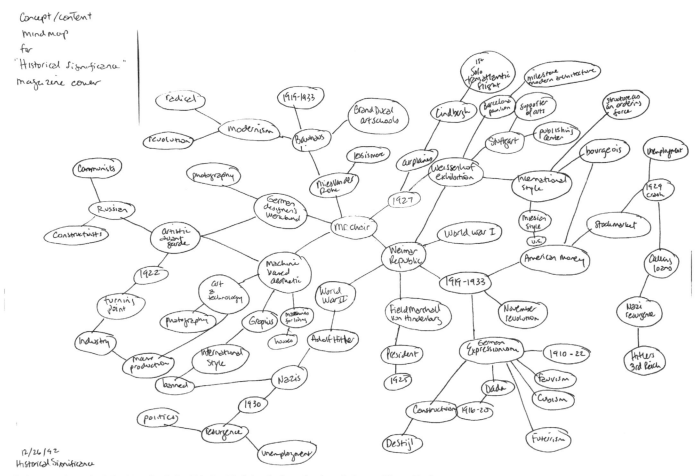

Concept/content
mind map
for
"Historical Significance"
magazine cover

12/26/92
Historical Significance

Mind maps about the Mies Van der Rohe Tubular Chair he designed using circles and lines. Each mind map explored a different central topic around the same theme. Models by erin malone.

HOW TO MAKE A MIND MAP

Making a mind map involves placing your central idea or topic in the center of the paper and writing down everything you can think of from that central topic.

List the subtopics around the main topic and use these to stimulate your brainstorming down that subtopic. Moving out from the center, write down ideas and related words one after the other in a train of thought until it is exhausted. Place a circle or oval around each idea and connect the terms with a line. This isn't required but I find it easier to follow the paths when the terms are separated from the connecting lines with a bounding shape. (I recommend staying away from squares to help differentiate from other kinds of maps.)

Each arm off the center radiates out, getting farther and farther away from the central topic. Most ideas won't connect from one leg of the diagram to another leg of the diagram but will be discrete trains of thoughts and in many cases, the outer ideas and concepts have absolutely nothing to do with the starting idea or with the concepts or ideas at the end of the other arms. That's ok.

Along the outer edges are the places you may find new ideas or elements or actors you may not have realized could be relevant to consider. Questions to ask yourself as you move down an idea path include what, where, when, why, how and who, as related to the central topic.

With mind maps you want to go back and layer in new ideas and connections to create a richer landscape if concepts. This will take a few passes to capture everything.

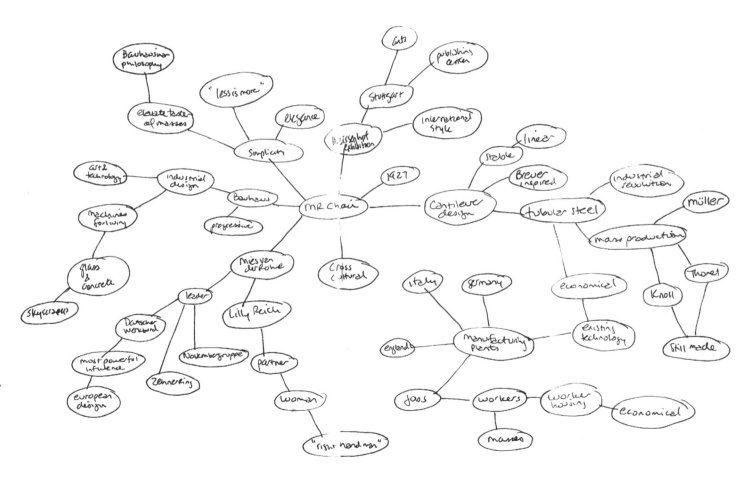

Mind map about the Mies Van der Rohe Tubular Chair he designed using circles and lines. Each mind map explored a different central topic around the same theme.
Model by erin malone.

GUITAR MIND MAP

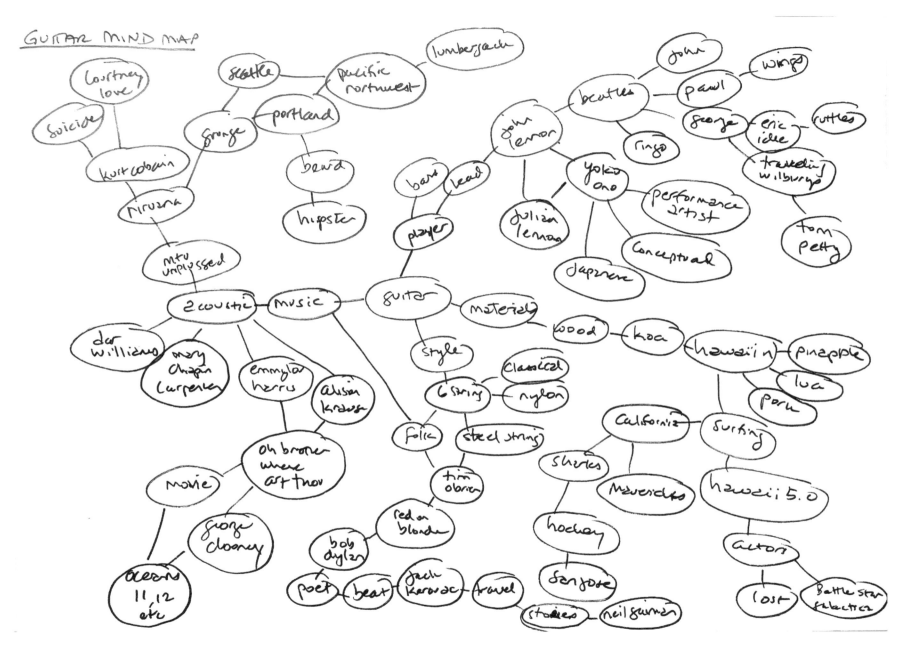

The Mind Map for Guitar—as a simple example showing train of thoughts to their logical end. Note how I end up with Lost, Neil Gaiman, Oceans 11 and Lumberjacks at the edges. Model by erin malone.

CLUSTER MAPS

WHAT IS IT?

The cluster map helps you organize your concepts into groupings and to start to see relationships between those clusters and your central topic. Visually, the cluster map looks similar to the mind map, but the difference between the two is that the content in a cluster map is all related to the central idea and is organized around sub-topics. The cluster map forces you to group ideas and terms together that are related along each arm of the diagram. Additionally, concepts may be connected across the radiating arms.

WHEN DO YOU DO ONE?

A cluster map can be made to help organize ideas, actors, agents, elements or components seen in a system or simply to help organizing ideas for a talk, essay or other writings. It is a visual outline with no hierarchy yet. The cluster map can aid you or your team in determining which areas are important to understand or drill deeper into and can provide the first stage of thinking about a system's architecture.

HOW TO MAKE A CLUSTER MAP

Making the cluster map starts the same way as the mind map. Put your main idea in the center of your page. Around the idea, place the next closely related sub-topics exploring elements, agents, actors, nodes, components.

Thinking about each of these sub-ideas, brainstorm all the related ideas and place them on the map going out from the central idea. Tease out non-obvious parts of the system and draw connections and relationships between the elements. If some words are related to others or are parallel, place these on the same level out from the center. As you move further out from the center, your ideas and topic may be several levels out from the sub-topics or main idea but unlike the mind map, they are all still related to the central concept.

Make your cluster map after you make the mind map, and use the mind map as a reference to make sure you capture all the related ideas. Once you have exhausted all the topics, components and elements, look at where there might be relationships across clusters and draw those connecting lines.

The final map may look more like a network map, depending on how related sub-topics are, than a mind map which may look more random and less organized.

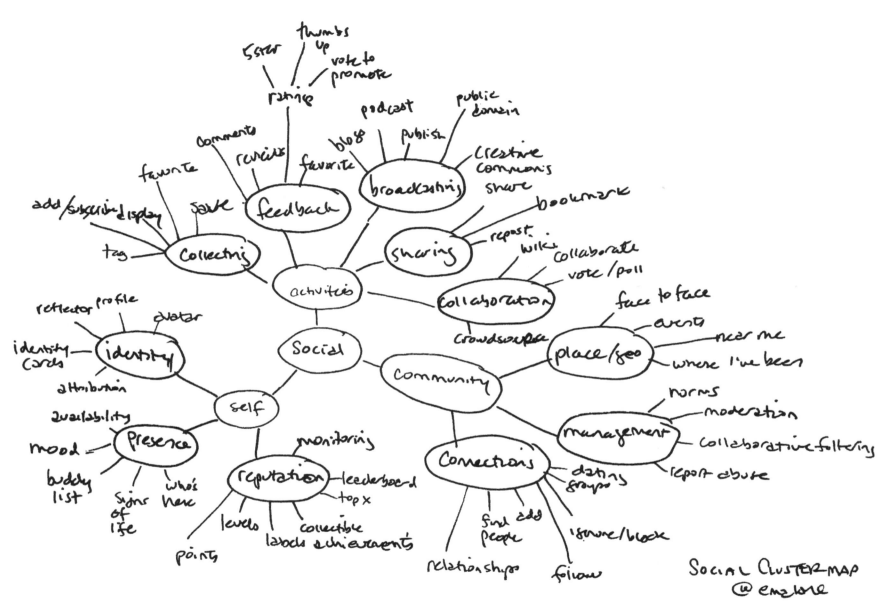

A Cluster Map for the Concept of Social. Note the clusters of related ideas, features, elements and actions coming off the term Social. Model by erin malone.

GUITAR CLUSTER MAP

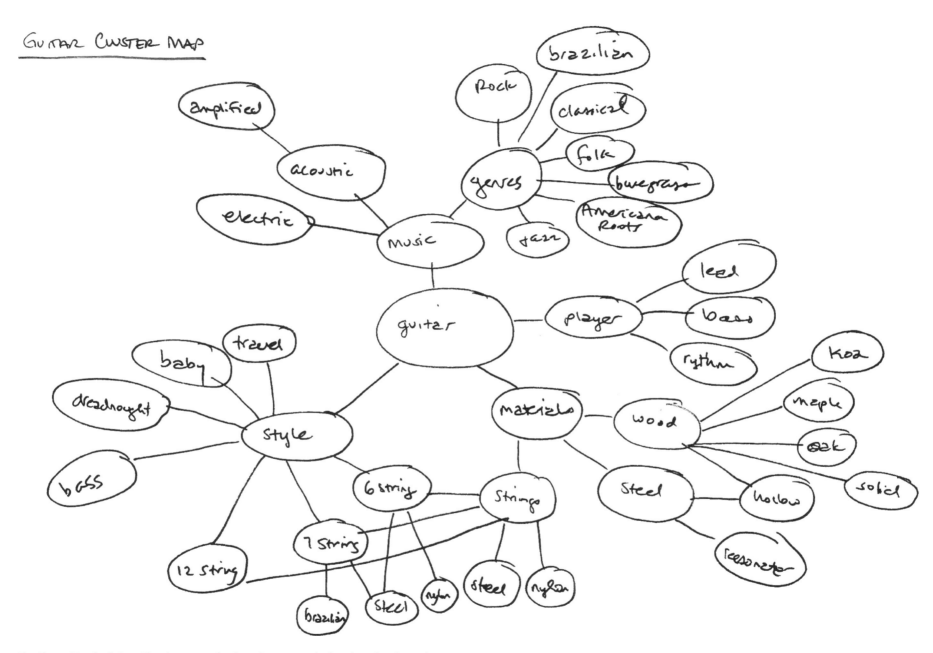

The Cluster Map for Guitar—The cluster map for the guitar starts to look at the subtopics and drills down along each of those to create groupings that are related to each other. Model by erin malone.

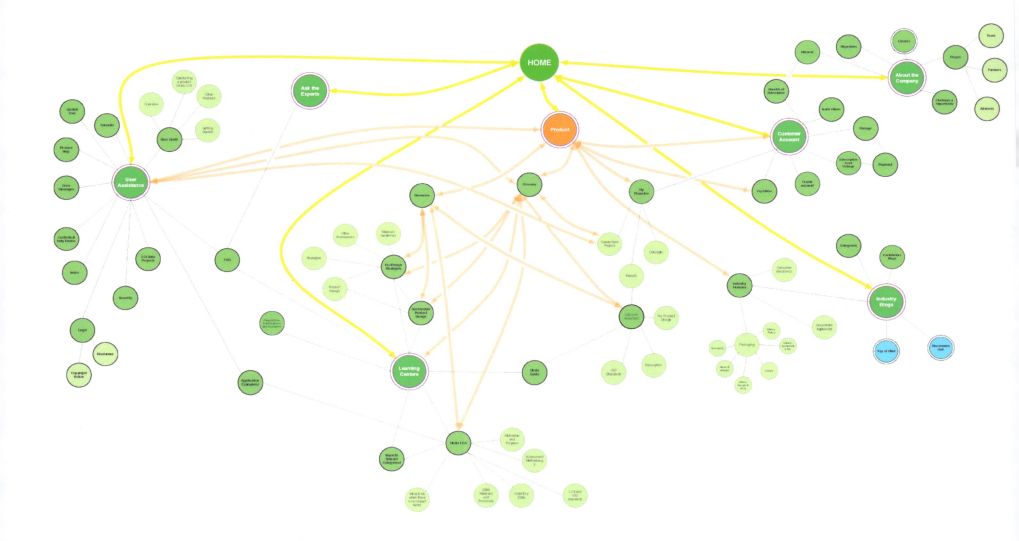

Cluster map exploring the different areas of a product's content categories. This was done prior to fleshing out the robust information architecture for the same product.

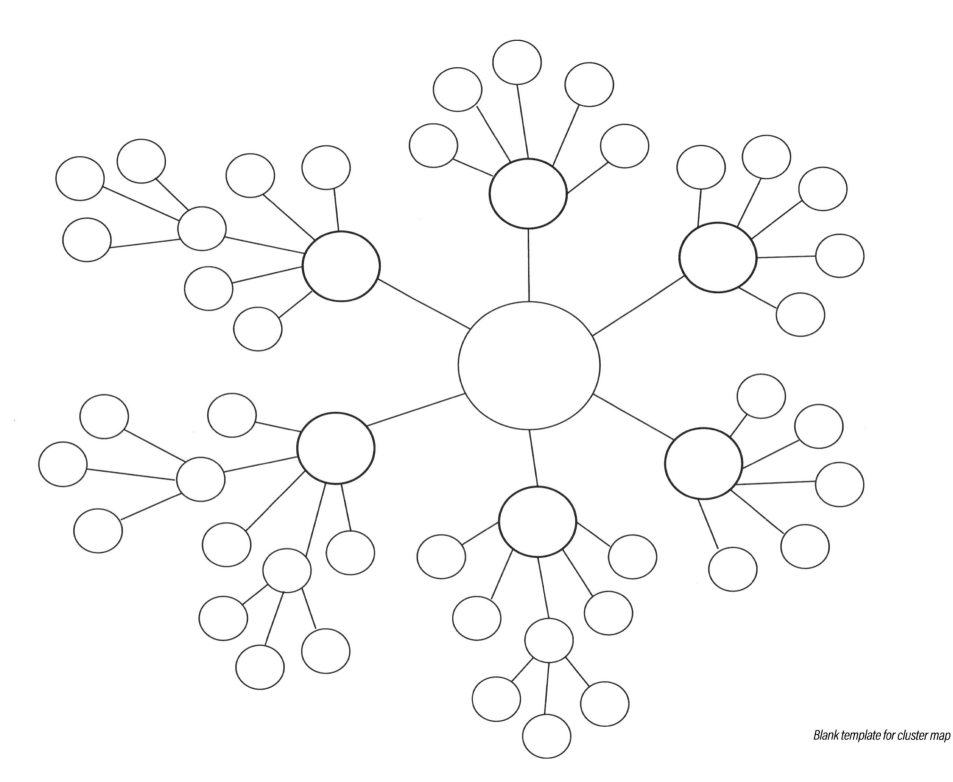

Blank template for cluster map

CONCEPT MAPS

WHAT IS IT?

The concept map, takes the elements, agents, actors and components fleshed out in the cluster map and organizes the ideas into a meaningful order that clearly articulates relationships between each item.

A concept map looks at the elements, agents, actors and components as nouns and the way they are related or connected as the verbs. These words can also be the bones of a data model later on in your project.

A well organized concept map can be read across, down and diagonally in the form of a sentences or a story that explains a train of concepts and their relationships to each other and the whole.

WHEN DO YOU DO ONE?

Early in the process to help understand the system, all the players - actors and data objects and entities, and to explore and define the relationships between all these elements within the system.

Hierarchy of concepts and entities may be revealed through this process. Interactions between entities may be revealed through this process.

I do concept maps a lot and follow the style and form laid out by Hugh Dubberly in much of his modeling work. (https://www.dubberly.com/concept-maps/creating-concept-maps.html)

HOW TO DO A CONCEPT MAP

To start a concept map, place the main idea at the top left of your page. Working to the right, look at what that item contains, what it uses, how it is related to, or is part of the next item. The verbs that generally show up in a concept map may include things like—which includes, is part of, can be, is, enabled by, with, should be, defined by, with the help of, contains, have, use, behaving as, represented by, to, called, a type of, etc.

Using the cluster map as a guide for related items (your nouns), work your way across exploring and completing a single coherent idea. Then from each of those main concepts (noun), work down or diagonally creating new ideas, exploring the nouns from your cluster map and the verbs that join them together. If you didn't do a cluster map, make a list of all the actors and objects in the system to start.

Each idea, agent, actor, component or element is placed in an oval. The ovals are connected by lines and the verbs sit on the lines between the concepts, connecting them and giving them context.

The concept map shows relationships between these ideas and between the threads through the connections and can help you or your team understand all the factors in play within a system. When determining where to pay attention, what things to build or where to make change, having an understanding of these relationships can help prioritize or group priorities together. Because the concepts are related and connected, we can see what might happen if an element or actor is changed or is missing.

GUITAR CONCEPTMAP

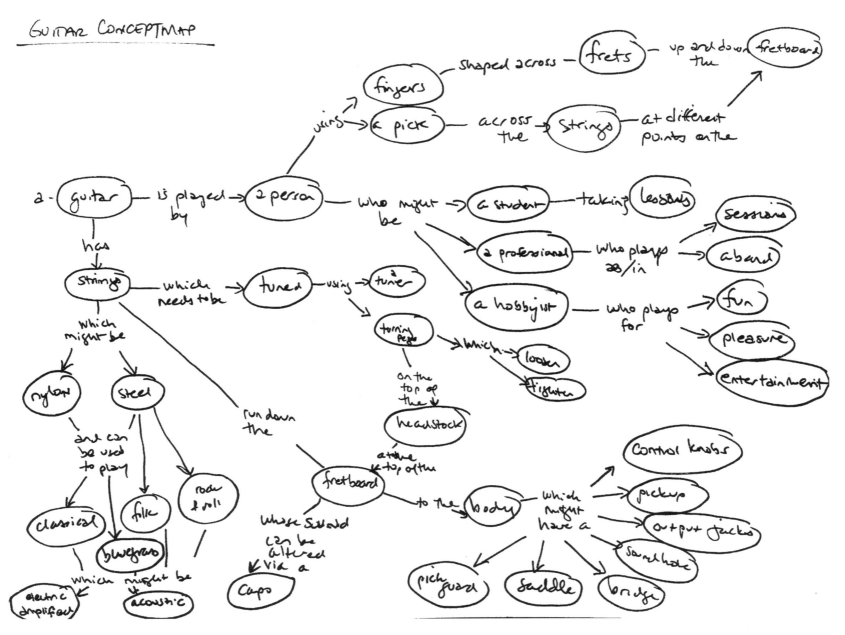

Rough draft concept map for Guitar. Concept maps generally reflect multiple iterations of exploration across the relationships and the list of actors to define the boundary. Model by erin malone.

Social Concept Map
@emalone

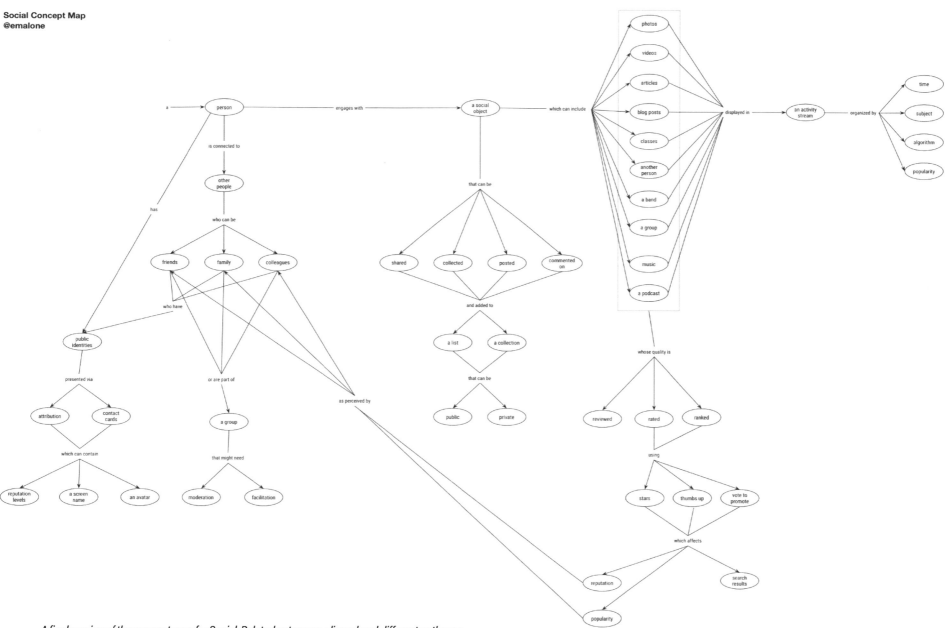

A final version of the concept map for Social. Related actors are aligned and different paths are explored off the main arm of the model at the top. Model by erin malone.

I usually start my Concept Maps with words. I write out sentences that explain or connect actors together through some sort of relationship.

For example:
An **object** can be given **tags** which can be organized into **collections**.

People can browse **collections**.

An **object** is posted by **a hunter**. The **object** can be found by **a finder** or **an expert** and featured on **the site**.

A **hunter** follows **an expert**.

An **expert** can be given **rewards** by **a hunter**.

Nouns are the actors and the verbs indicate how the different actors are related. When diagramming, the nouns go in ovals and the verbs are part of the connecting lines.

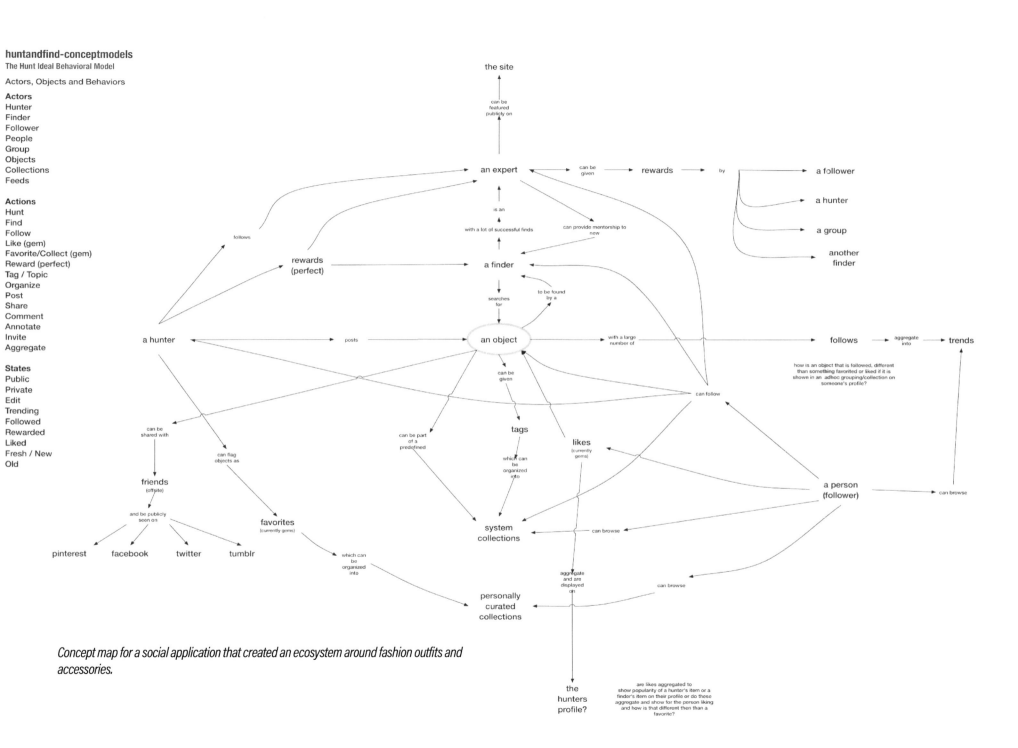

huntandfind-conceptmodels
The Hunt Ideal Behavioral Model

Actors, Objects and Behaviors

Actors
Hunter
Finder
Follower
People
Group
Objects
Collections
Feeds

Actions
Hunt
Find
Follow
Like (gem)
Favorite/Collect (gem)
Reward (perfect)
Tag / Topic
Organize
Post
Share
Comment
Annotate
Invite
Aggregate

States
Public
Private
Edit
Trending
Followed
Rewarded
Liked
Fresh / New
Old

the site

an expert — can be given → rewards — by → a follower / a hunter / a group / another finder

is an
with a lot of successful finds
can provide mentorship to new

follows

rewards (perfect) → a finder

searches for
to be found by a

a hunter — posts → an object — with a large number of → follows — aggregate into → trends

can follow

how is an object that is followed, different than something favorited or liked if it is shown in an adhoc grouping/collection on someone's profile?

can be given

can be shared with

can flag objects as

can be part of a predefined

tags
which can be organized into

likes (currently gems)

a person (follower)
can browse

friends (offsite)

and be publicly seen on

favorites (currently gems)

which can be organized into

pinterest   facebook   twitter   tumblr

system collections
can browse

aggregate and are displayed on
can browse

personally curated collections

the hunters profile?

are likes aggregated to show popularity of a hunter's item or a finder's item on their profile or do these aggregate and show for the person liking and how is that different then than a favorite?

Concept map for a social application that created an ecosystem around fashion outfits and accessories.

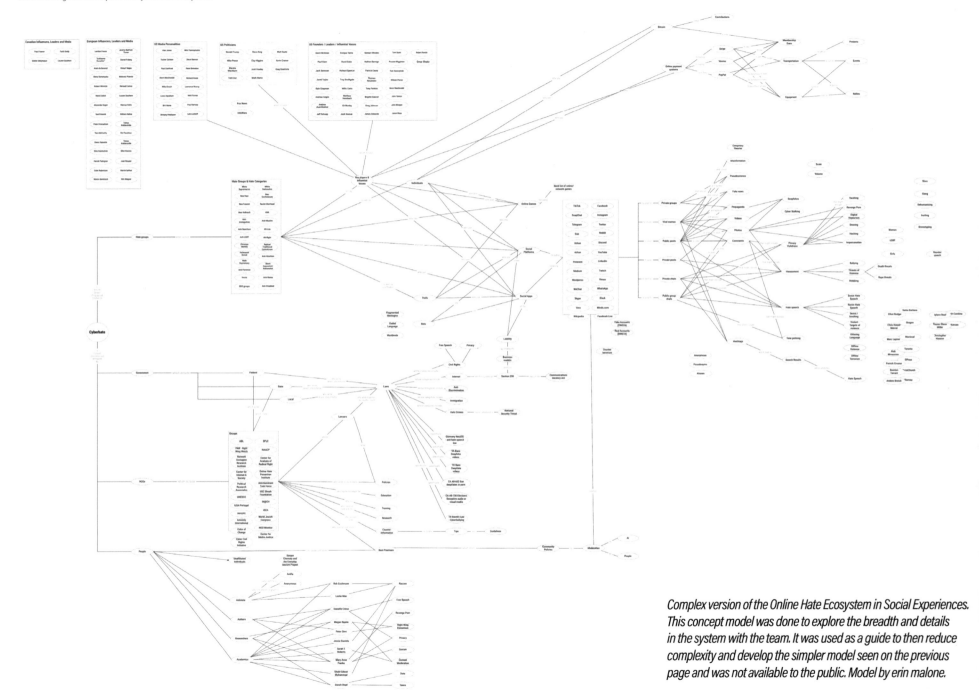

Complex version of the Online Hate Ecosystem in Social Experiences. This concept model was done to explore the breadth and details in the system with the team. It was used as a guide to then reduce complexity and develop the simpler model seen on the previous page and was not available to the public. Model by erin malone.

Relationships in the Cyber Hate
Ecosystem Between the Main Actors

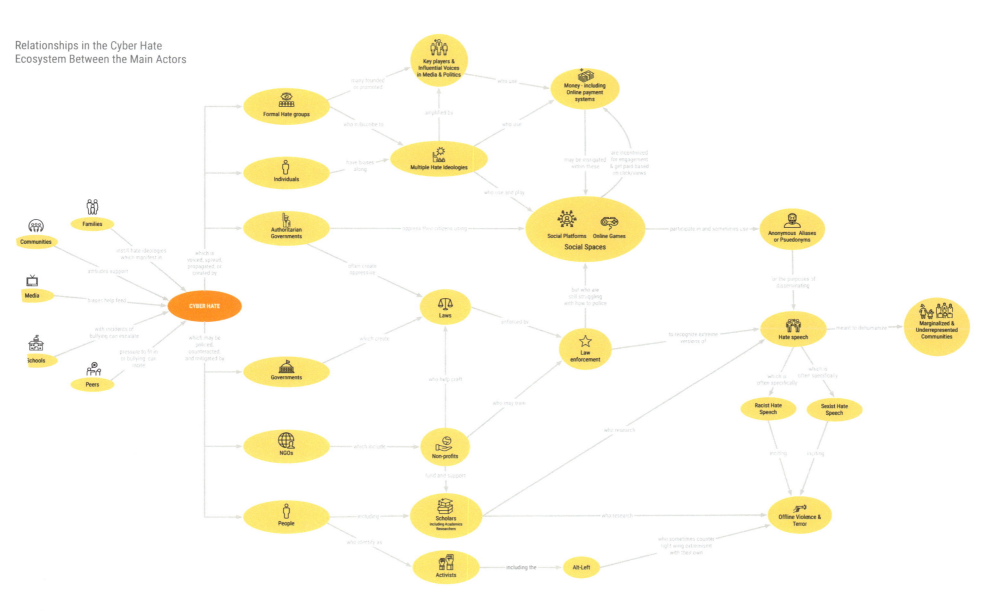

The relationships between actors in the Online Hate Ecosystem is a simplified concept model used to visualize this system for the Anti Defamation League. Model by erin malone

STOCK FLOW DIAGRAMS

WHAT IS IT?

Stock flow diagrams capture the primary entities within a system and the factors and elements that affect them, either in a positive or negative way, thereby creating reinforcing or balancing feedback loops within the system.

WHEN DO YOU DO ONE?

This type of model can be done at the beginning of a project to understand the larger ecosystem within which a product or service might play within. The models can also be done to help understand how users might behave within a system and factors and points of leverage can influence those behaviors.

Stock Flow diagrams are unusual in the UX design space but as we become more and more involved in solving wicked problems and bringing UX design approaches to these problems, traditional systems thinking, as taught and written about by Donella Meadows in her book Thinking In Systems, is appropriate to get our heads around these dynamic and ever changing ecosystems.

HOW TO DO A STOCK FLOW DIAGRAM

A stock flow diagram shows interdependencies and feedback within a system by identifying major accumulations and the elements and actions that cause an increase and decrease of those accumulations over time.

Anything within the system that can be accumulated is a Stock. In the case of a social ecosystem, or reputation, this might be things like posts or articles read or page views or time on site or a thousand other things we see in digital experiences.

It can also be an abstract concept like happiness or love. Generally stocks are Nouns which can be both concrete and abstract. To best identify which stocks to include in a diagram, we need to analyze the critical behaviors of the system. We might do this through Concept Models to understand relationships or Behavior Over Time graphs to see changes over time.

A stock may have an input flow (increase) or an output flow (decrease) or both. A flow can have multiple influences and can have stacked influences, i.e. one element can influence another which then can influence the flow to increase or decrease.

The Flows in and out are represented by an arrow into the Stock and an arrow out of the Stock. On each arrow there is a round circle with a lever on top. This symbol represents the levers that turn up or turn down the flow in

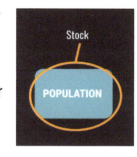

Stock for Earth's Population. Population is something that can be accumulated and can increase and decrease with different influences. In the model, we represent the stock with by a box.

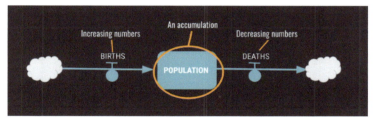

Basic stock flow diagram showing inflows and out flows for the stock Population. The clouds represent boundaries – outside the scope of the system we are analyzing.

Input flow with connectors into the flow that INCREASE the idea of Good Online Reputation.

Elements that help increase online reputation include quality of posts, the number of followers, the number of posts, the number of likes on those posts and the number of reposts. In aggregate, an algorithm of these factors give other people clues to the reputation for a person online.

Output flow with connectors into the flow that cause a DECREASE in the idea of Good Online Reputation.

Notice, that in some cases, less of something will cause the outflow as well as more of something influencing the outflow.

In this model, more spam can increase the OUTFLOW or DECREASE Good Online Reputation. It damages that reputation. But also a DECREASE in post quality can also increase the decline of Good Online Reputation. The nuances implied by the pluses and minuses are important to understanding cause and effect and that accumulation in the stock.

and out of the stock. Like the hot and cold controls of a faucet. The elements that influence the flow, Connectors, indicate that changes in one element cause changes in another element.

Elements Increasing Reputation
Stock Flow Diagram
@emalone

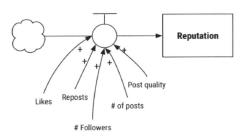

Conceptually, these increase or decrease the flow. (Stocks can only be affected through flows, therefore you should not attach a connector directly to a stock.)

Elements Decreasing Reputation
Stock Flow Diagram
@emalone

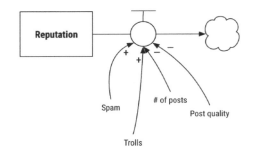

Using the example of Reputation in social experiences, if we write this out in a narrative form it reads something like this:

An increase in posts, likes, quality, reposts, can increase (the input flow) a person's online reputation within the system (the stock). An increase in self-promotion, ads, spam, or trolls in the comments, as well as a decrease of posts or decrease in quality of those posts, can decrease (the output flow) a person's reputation within the system (the stock).

Writing it out can help clarify how this idea of reputation works but it is hard to visualize these influences in text and it's a mouthful to describe.

Frankly, my eyes glaze over a bit when trying to read it and understand what is happening. We can draw it out and visually understand the ins and outs and influences to the stock, reputation.

See the two examples on the previously page.

Social Media Influence
Stock Flow Diagram
@emalone

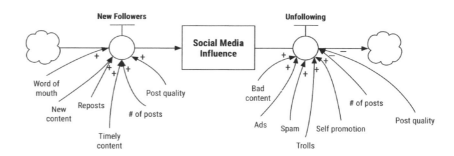

**Reputation System
Stock Flow Diagram**
@emalone

Once you have done a simple stock flow diagram about an area in a system, you can start to identify other stocks in the overall larger system. Analyzing these in the same way, we can look at how these may be interconnected and influence each other and we may start to see the emergence of some feedback loops.

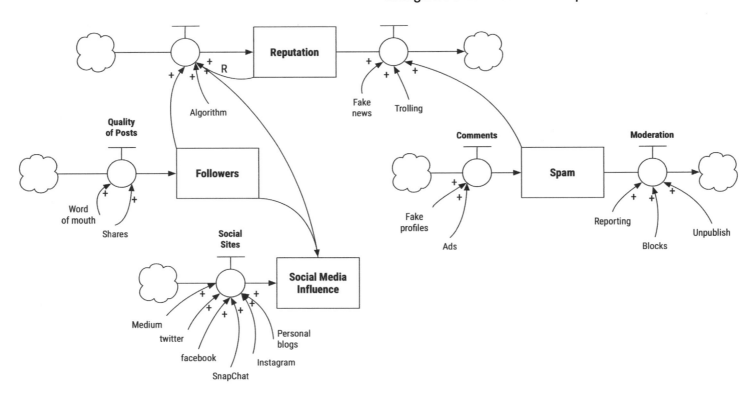

A more complex diagram of stocks and flows that influence Reputation. There are multiple stocks in the system and they can influence each other in various ways. Creating and making the stock flow model helps you understand some of the influential relationships between elements in the system. Model by erin malone.

SITE MAPS

WHAT IS IT?

A sitemap captures all sections of a website or application and usually presents the topics or task types in a hierarchical manner. There are other models besides hierarchical, including hub and spoke and dynamic - where all pages connect to all pages.

Usually each block in the diagram represents a "page" or "screen" and identifies a template that must be built by the development team. Sometimes more detailed sitemaps include sections documenting tasks but this gets complicated in terms of defining the templates needed. Subtasks or screen parts may be included and are usually identified through some graphical differentiation like color or dotted lines or an inventory number so that each component up to the top level page template is captured.

In more simple sites and many that I have created, the sitemap is hierarchical and doesn't show cross linking or elements that are shared horizontally in a system. Sitemaps shouldn't replace user task flows as they don't show system decisions and responses.

WHEN DO YOU DO ONE?

This type of model is done early on in the process and is updated and edited as a project is defined. Initially, it can be done to help define scope and understanding of the ecosystem at play. As the work progresses, the sitemap gets more refined, complete and layered with more details. A sitemap usually accompanies wire frames and/or interaction flow diagrams.

HOW TO MAKE A SITEMAP

Making a sitemap starts with some understanding of the scope of the site or application.

Generally, I start by sketching it out and thinking about the major sections that might be represented by navigation. And then I think about all the different screens I might need to have—things like settings, and help, and search results.

I make a list on paper or index cards and then organize the cards the way I think the site should be organized. For super large projects, post it notes on the wall work. With cards and post it notes, you can rearrange as needed when the organization needs to change.

Then I start mapping. Starting with the top level screen first - usually the home page. All other screens or pages cascade off that, with each being a separate rectangle. You will see that sitemaps look like fancy org chart. Because in most cases they are hierarchical.

With content rich sites, metadata and facets can be shown as a list next to the search results box, or next to a details page box.

There are multiple ways to display that extra information depending on complexity. Sometimes its not there at all on the map, but captured in a spreadsheet because there is too much to visualize easily.

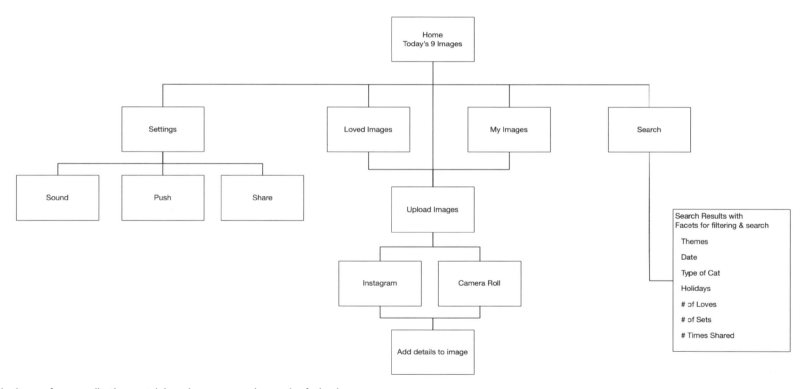

Super simple sitemap for an application containing a homepage and a couple of other lower level pages. Facets for the search results filters are shown in a list.

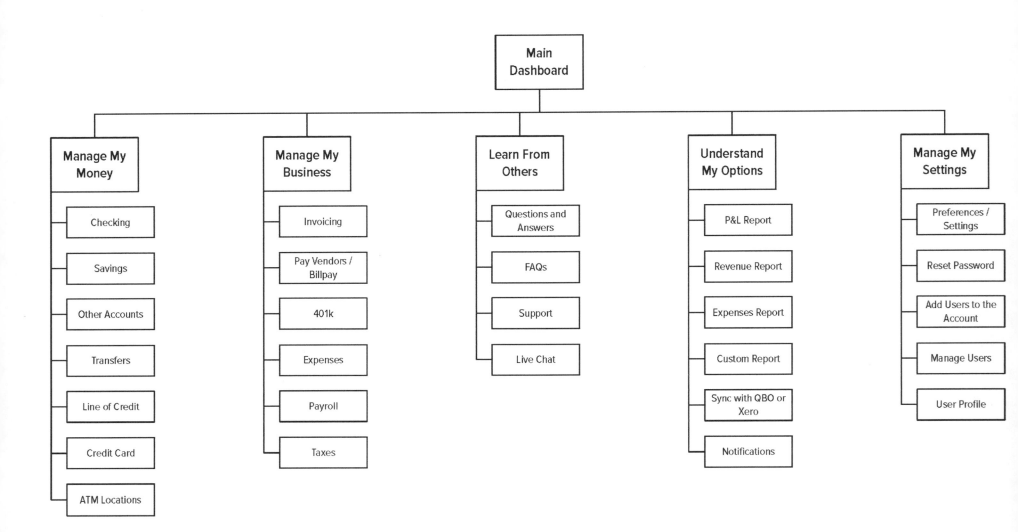

```
                              ┌──────────────┐
                              │     Main     │
                              │  Dashboard   │
                              └──────────────┘
```

Main Dashboard

- **Manage My Money**
  - Checking
  - Savings
  - Other Accounts
  - Transfers
  - Line of Credit
  - Credit Card
  - ATM Locations

- **Manage My Business**
  - Invoicing
  - Pay Vendors / Billpay
  - 401k
  - Expenses
  - Payroll
  - Taxes

- **Learn From Others**
  - Questions and Answers
  - FAQs
  - Support
  - Live Chat

- **Understand My Options**
  - P&L Report
  - Revenue Report
  - Expenses Report
  - Custom Report
  - Sync with QBO or Xero
  - Notifications

- **Manage My Settings**
  - Preferences / Settings
  - Reset Password
  - Add Users to the Account
  - Manage Users
  - User Profile

A fairly straightforward and simple architecture for a banking product. The sitemap doesn't show any facets or any kind of indication of micro-interactions or content blocks to be included in the pages. This was a first pass. Later iterations would most likely include those other elements. Map by erin malone.

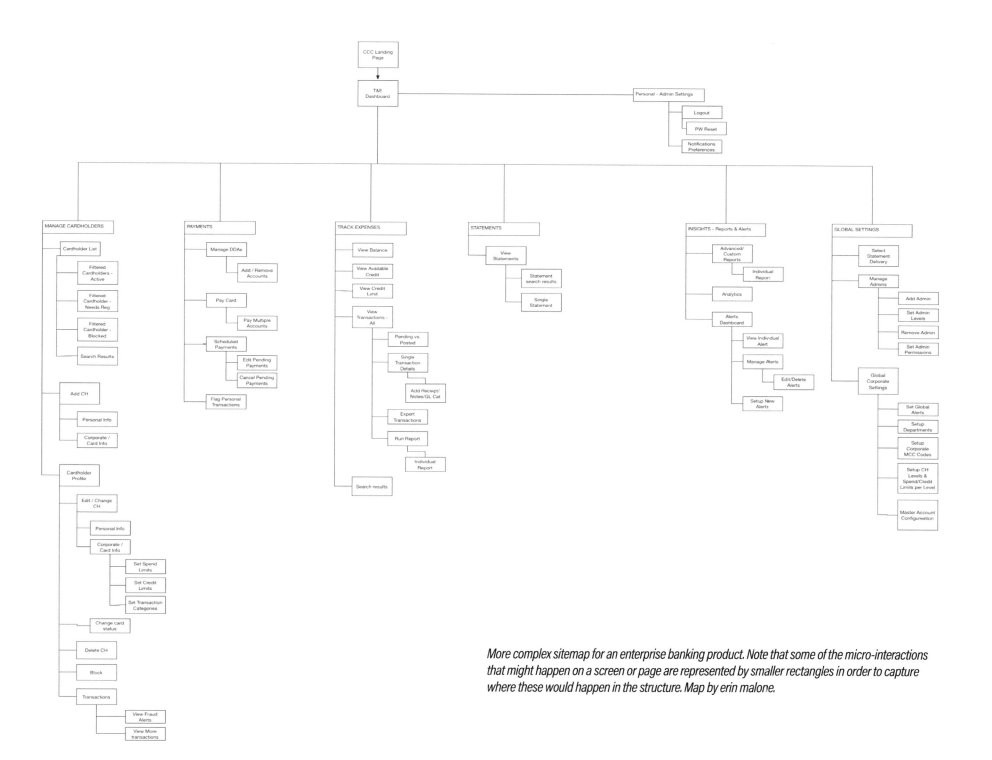

More complex sitemap for an enterprise banking product. Note that some of the micro-interactions that might happen on a screen or page are represented by smaller rectangles in order to capture where these would happen in the structure. Map by erin malone.

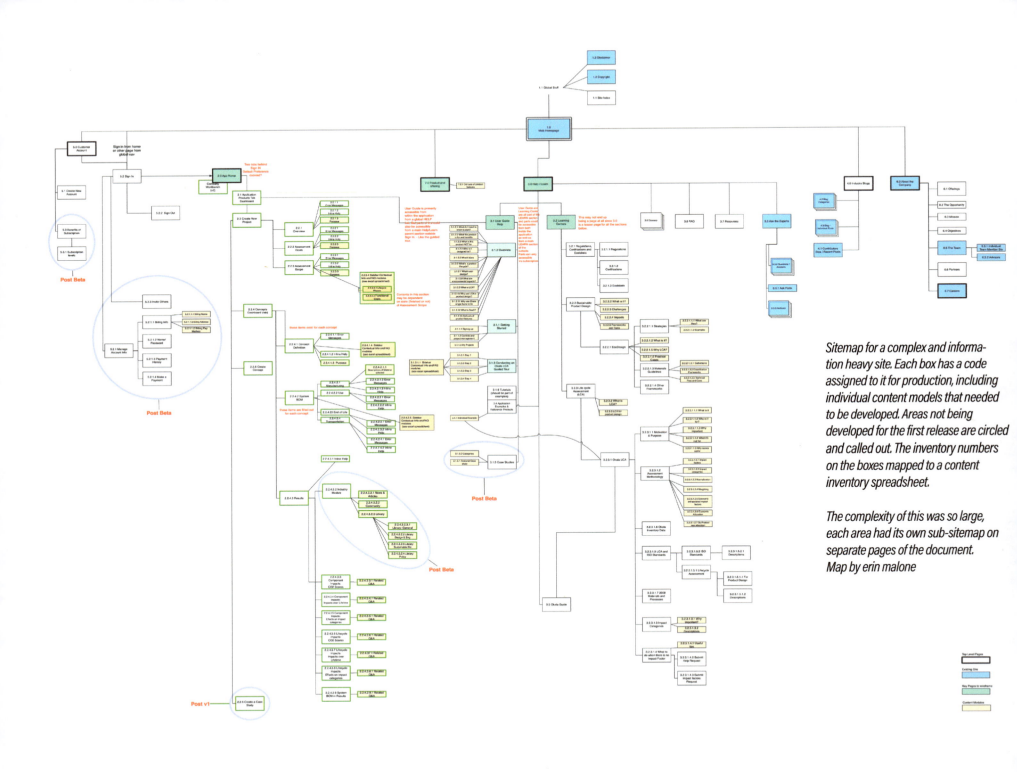

Sitemap for a complex and information-heavy site. Each box has a code assigned to it for production, including individual content models that needed to be developed. Areas not being developed for the first release are circled and called out. The inventory numbers on the boxes mapped to a content inventory spreadsheet.

The complexity of this was so large, each area had its own sub-sitemap on separate pages of the document. Map by erin malone

TASK & TIME TARGET MODEL

WHAT IS IT?

A diagram that expresses the types of tasks a user may need to do, want to do or should be doing intersected with time which starts with the smallest increments in the middle and durations or frequency lengths grow towards the outer rings of the circle.

The target is divided up into sections to help group the elements and concepts and in the example below, the two halves of the circle represent internal motivations, attitudes and desires and external interactions and relationships. The tasks are organized accordingly.

WHEN DO YOU DO ONE?

Early in the process to help understand the system, all the players - actors and tasks and how they relate to each other through time. Additional information may be layered in by dividing up the target and clustering like type tasks or entities together.

Ecosystem - Concept Model

Horizon Lens

FAST TASKS
Emergency and one off tasks

Daily

Weekly or Bi-weekly

Monthly

Quarterly

Yearly

INTERNAL
Attitudes & Desires

Stay informed

Have peace of mind

EXTERNAL
Interactions & Relationships

Keep business happy
(i.e. owner & IRS)

Keep employees happy

Horizon - solar system view

Activities and tasks revolve around the business owner in an elliptical shape and are organized by time or frequency (i.e. things done daily versus monthly) with the center being the most frequent—daily—and the outer ring being the least frequent—yearly.

Business Owner circle: As tasks cross into the horizon or orbit - i.e. End of december - paying bills, paying employees, revamping 401k and insurance, reviewing P&L for year, calling clients - there is a mix of daily, monthly, quarterly and yearly tasks being attended to.

Tasks and activities that address internal attitudes and desires for the business and/or user are clustered towards the left of their respective ellipse.

Tasks and activities that involve external actors in the business ecosystem are clustered towards the right of their respective ellipse.

Be efficient

Make money

INTERNAL
Attitudes & Desires

Keep clients happy

Keep vendors happy

EXTERNAL
Interactions & Relationships

Early iteration of the ecosystem model where the main user is at the left and the tasks organized by time revolve around them in an oval shape. Model by erin malone.

Ecosystem - Concept Model

Cadence Lens with Horizon Alley - CO Tasks

Be efficient

Keep business happy (i.e. owners & IRS)

Yearly
Quarterly
Monthly
Semi-monthly
Weekly
Daily

Prepare business K1s
P&L for the year
Prepare business tax taxes
Get line of credit
Form 5500 for 401k
Statement of Information to State

Close out books
Review P&L forecast
Pay Payroll taxes
Pay Gross Receipts taxes

Open a line of credit
Balance books
Pay Payroll taxes
Plan yearly to-do's for biz
Pay quarterly sales tax
Pay quarterly estimated taxes

Order checks
Set up auto-pay for bills
Pay down line of credit
Send 1099's
Update 401k offerings

Pay business insurance premiums
Collate monthly timesheets
Send financials to CPA for review
Plan quarterly to-do's for biz
Approve expense reports
Set up Payroll for new employees
Give employees raises

Have peace of mind
Keep employees happy

Confirmation that deposits were received
Transfer money between accounts
Reconcile transactions
Review P&L forecast
Plan monthly to-do's for biz
Pay employees
Set up Direct Deposit for employees
Update insurance offerings

Review disability insurance
Check if payments cleared
Upload receipts - expenses
Review trending numbers
Submit 401k deposits
Create expense reports

Reconcile transactions
Plan weekly to-do's for biz
Review weekly timesheets

Transfer money from line of credit
Check balances on accounts
Review bank balances
Interact with employees

INTERNAL Attitudes & Desires

Daily

EXTERNAL Interactions & Relationships

Review office space lease
Accept Square/PayPal/SPARKPay payment
Pay bounced check or late bill

Pay bounced check or late bill

Accept payments against invoices
Make deposits
Review daily timesheets
Calls with customers
Visits with customers
Send out invoices
Give financial - project updates to clients

Read notifications
Talk to vendors
Bring on new customers
Give financial updates to clients
Make notes about client specifics

Make deposits
Review weekly timesheets
Read important bank emails
Meet new vendors
Give financial - project updates to clients
Set up Direct deposit from clients

Make money
Keep clients happy

Collate monthly timesheets
Review cashflow
Review end of month to-do's
Send orders to vendors
Set up Direct deposit from clients

Review line of credit
Review credit card charges
Pay Bills - Vendors
Send W9 to new clients

Review revenue by client
Review monthly reports
Pay Bills - Utilities

Review revenue by client
Review revenue by client
Review expenses by vendor
Set up Direct deposit for vendors
Get W9 from new vendors

Stay informed
Review expenses by vendor

Keep vendors happy

Business owner centric view

Activities and tasks revolve around the business owner and are organized by time or frequency (i.e. things done daily versus monthly) with the center being most frequent—daily— and the outer ring being the least frequent —yearly

Center arm: At any point in time, the business owner might be doing an activity or task across the spectrum of time. i.e. End of december - paying bills, paying employees, revamping 401k and insurance, reviewing P&L for year, calling clients - so at any point there is a mix of daily, monthly, quarterly and yearly tasks being attended to.

Left side: Tasks and activities that address internal attitudes and desires for the business and/or user

Right side: Tasks and activities that involve external actors in the business ecosystem

---

Diagram to work through possible tasks by time and which stakeholders would care. Each possible item that a product could support is listed and organized by time and whether it meets internal or external needs. This might be put together after user research insights are developed capturing user needs and motivations. Model by erin malone.

TAXONOMY DIAGRAM

WHAT IS IT?

A diagram that captures all the categories, subcategories and potential metadata (tags and keywords) associated with a topic, a site, an application or a service. The model can take the form of a diagram (like below) or as a spreadsheet or system of spreadsheets.

WHEN DO YOU DO ONE?

The taxonomy can be started as soon as a project begins to help define the scope of understanding in terms of information categorization and organization. The taxonomy tends to be a living document and is usually updated and edited throughout the design process and beyond as part of an overarching content strategy and governance system.

Many people capture their taxonomies in spreadsheets. In the case of these examples, the scope and categories needed to used for communication with external stakeholders and a spreadsheet wasn't the right vehicle.

Example on the following page:
Social Ecosystem Diagram - shows the taxonomy of the social interface patterns developed and written in the book Designing Social Interfaces by Christian Crumlish and Erin Malone. The taxonomy was a result of multiple sessions of diagramming with the community, card sorting exercises and decisions around the book structure. Inspired by models by Nancy Duarte. Model by erin malone.

Social Spaces

Activities
Self
Community

Engagement
Sign Up / Register
Sign In
Sign In Continuity
Sign Out
Receive Invitation
Send Invitation
Authorize
Private Beta
Welcome Area
Reengagement

Identity
Identity
Profile
Profile Decorating
Reflectors
Identity Cards
Attribution
Testimonials
Avatar
Personal Dashboard

Collecting
Saving
Favorites
Displaying
Add / Subscribe
Tag an Object
Find with Tags
Tag Cloud

Feedback
Comments
Reviews
Ratings
Favorites
Vote to promote
Thumbs up/down ratings
Solicited Feedback

Broadcasting
Publishing
Blogs - Read
Blogs - Publish
Microblogging
Publishing
Rights
Terms of Service
Licensing

Sharing
Sharing
Bookmarklet
Share This
Activity Streams
Send This
Casual Privacy
Give Gift
Many Publics
Social Bookmarking
Embedding
Ongoing

Communicating
Synchronous vs. Asynchronous
Sign In to Participate
Forums
Public Conversation
Private Conversation
Nudging
Group Conversation
Flamewars
Vendettas
Sock-Puppets

Collaboration
Collaboration
Manage Project
Voting
Collaborative Editing
Edit this page
The Wiki Way
Unlock
Crowdsourcing

Social Media
Following
Filtering
Recommendations
Social Search
Pivoting
Monetize
Underparticipation

Presence
Availability
Mood
Environment
Buddy List
Statuscasting
Microblogging
Updates Opt-in
Signs of Life
Usergallery
Who's Here Now
Ambient Intimacy

Reputation
Levels
Labels
Collectible Achievements
Temporal Awards
Peer-to-peer Awards
Points
Leaderboard
Top X
Statistical Evidence
Monitoring
Friend Ranking

Connections
Relationships
Find People
Add Friends
Implicit / Explicit Relationships
Fans & Fame
Circles of Connections
Publicize Relationships
Unfriend

Community
Management
Norms
Manifesting
Collective Choices
Group Moderation
Collaborative Filtering
Report Abuse
What's the Story

Place
Geo
Location
Face-face Meeting
Party
Calendaring
Reminding
Geo-Tagging
Geo-Mapping
Geo-Mashing
Neighborhood
Mobile - Geo
Mobile - Gatherings
Mobile - Statuscasting

Respect the Ethical Dimension
Don't Break Email
Learn From Games
Be Open
Password Anti-Pattern
Cargo Cult Anti-Pattern
ExBoyfriend Anti-Pattern
Pumpkin Village

The Online Hate Perpetrators: Targets: Amplifiers and Mitigators diagram came through multiple sessions with stakeholders, rounds of diagramming concept models and multiple feedback sessions across a wide range of disciplines.
Model by erin malone.

Hear more about how this came about in this talk from Interaction21.
https://vimeo.com/510256429

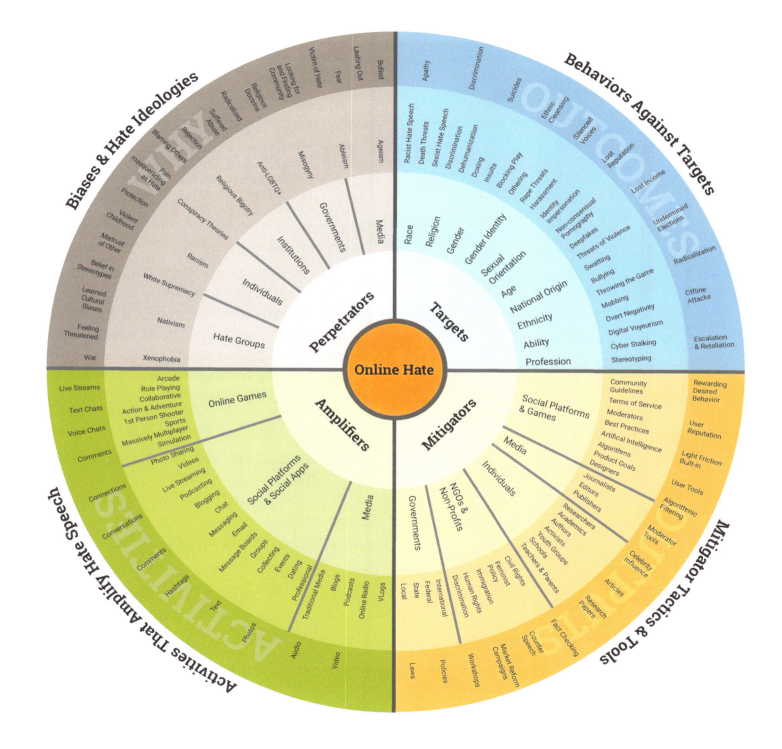

Online Hate:
Perpetrators
Targets
Amplifiers
& Mitigators

Biases & Hate Ideologies

WHY

Looking for and Finding Community
Religious Doctrine
Victim of Hate
Radicalized
Lashing Out
Suffered Abuse
Fear
Rejection
Bullied
Blaming Others
masquerading as Hate
Pain
Protection
Violent Childhood
Mistrust of Other
Belief in Stereotypes
Learned Cultural Biases
Feeling Threatened
War

Conspiracy Theories
Religious Bigotry
Anti-LGBTQ+
Misogyny
Ableism
Ageism
Racism
White Supremacy
Nativism
Xenophobia

Media
Governments
Institutions
Individuals
Hate Groups

Perpetrators

Online Hate

Targets

Race
Religion
Gender
Gender Identity
Sexual Orientation
Age
National Origin
Ethnicity
Ability
Profession

Behaviors Against Targets

OUTCOMES

Apathy
Discrimination
Suicides
Ethnic Cleansing
Silenced Voices
Lost Reputation
Lost Income
Undermined Elections
Radicalization
Offline Attacks
Escalation & Retaliation

Racist Hate Speech
Death Threats
Sexist Hate Speech
Discrimination
Dehumanization
Doxing
Insults
Blocking Play
Othering
Rape Threats
Harassment
Identity Impersonation
Non-consensual Pornography
Deepfakes
Threats of Violence
Swatting
Bullying
Throwing the Game
Mobbing
Overt Negativity
Digital Voyeurism
Cyber Stalking
Stereotyping

Amplifiers

Mitigators

Arcade
Role Playing
Collaborative
Action & Adventure
1st Person Shooter
Sports
Massively Multiplayer
Simulation

Live Streams
Text Chats
Voice Chats
Comments

Online Games

Photo Sharing
Videos
Live Streaming
Podcasting
Blogging
Chat
Messaging
Email
Message Boards
Groups
Collecting
Events
Dating

Social Platforms & Social Apps

Connections
Conversations
Comments
Hashtags
Text
Photos
Audio

Media

Professional
Traditional Media
Blogs
Podcasts
Online Radio
Vlogs

Video

Activities That Amplify Hate Speech

ACTIVITIES

Social Platforms & Games

Community Guidelines
Terms of Service
Moderators
Best Practices
Artifical Intelligence
Algorithms
Product Goals
Designers

Rewarding Desired Behavior
User Reputation
Light Friction Built-in
User Tools
Algorithmic Filtering
Moderator Tools
Celebrity Influence
Articles
Research Papers
Fact Checking

Media

Journalists
Editors
Publishers

Individuals

Researchers
Academics
Authors
Activists
Youth Groups
Schools
Teachers & Parents

NGOs & Non-Profits

Civil Rights
Feminist Policy
Immigration
Human Rights
Discrimination

Governments

International
Federal
State
Local

Laws
Policies
Workshops
Market Reform Campaigns
Counter Speech

Mitigator Tactics & Tools

TACTICS

CUSTOMER JOURNEY & EXPERIENCE MAPS

WHAT IS IT?

User journeys and experience maps are often created to help understand how a customer or user moves through a multi-touch point experience. The map expresses the flow across modes, the main tasks the customer is trying to achieve, their goals, emotions, hopes and desires during this journey and their pain points that are encountered along the way.

A journey map can be started with post-it notes on a wall and then refined over time as a digital artifact.

WHEN DO YOU DO ONE?

This type of model is done as part of user research and ethnography. Journeys can be used to support the archetypal experiences of a persona in the system and helps reveal areas for improvement as well as opportunities that might not otherwise have been discovered.

They often reflect the primary user scenarios in the service.

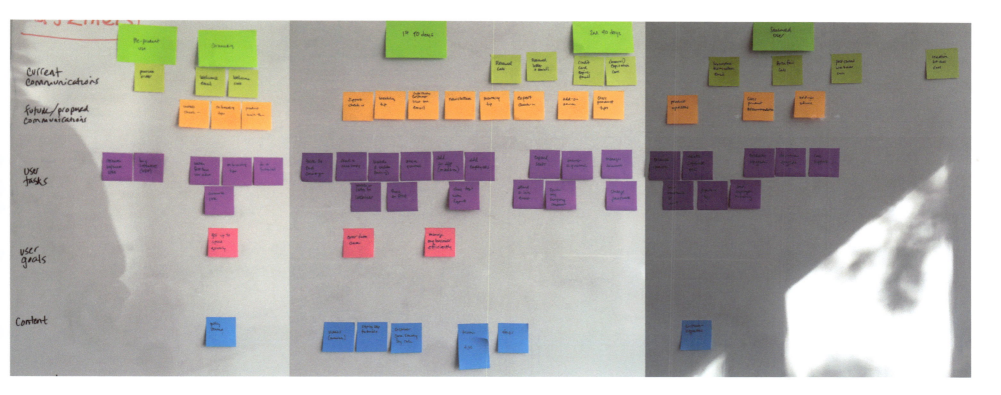

HOW TO MAKE A JOURNEY MAP

Do a lot of user research, ethnography and user interviews, stakeholder interviews and if a product already exists, an audit of the system and its interactions. Also consider talking with customer support representatives and analyzing the analytics for the site or service.

Then breakdown the journey into tracks— tasks, goals, user's hopes and their emotions while experiencing the product or service.

Capturing negative emotions and potential areas that trigger customer support calls identifies areas of opportunity for design or service improvements.

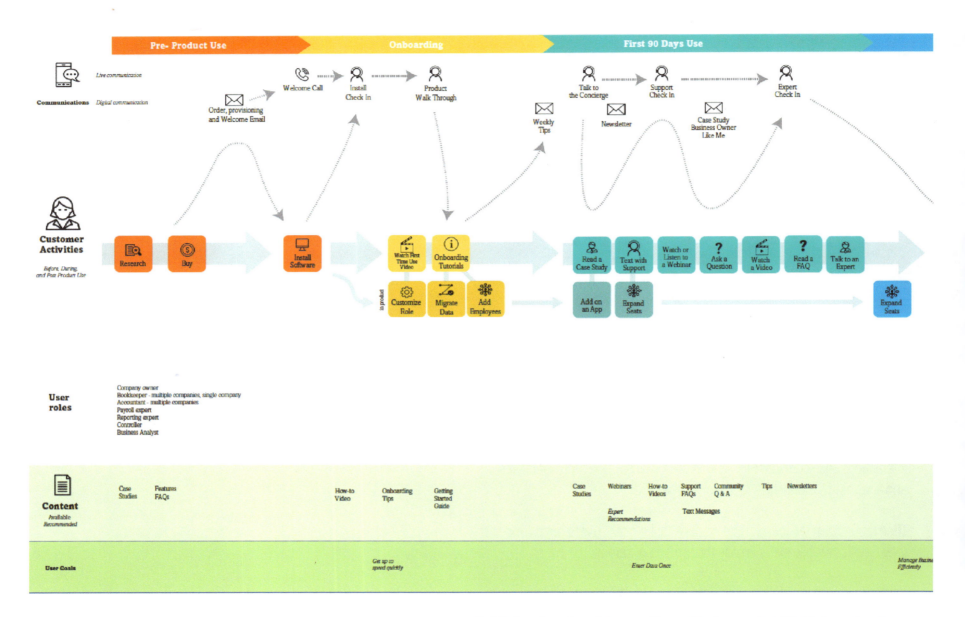

Communications

Live communication

Digital communication

Order, provisioning and Welcome Email

Welcome Call
Install Check In
Product Walk Through

Talk to the Concierge
Support Check In
Expert Check In

Weekly Tips
Newsletter
Case Study Business Owner Like Me

Pre- Product Use   **Onboarding**   **First 90 Days Use**

Customer Activities

Before, During, and Post Product Use

Research
Buy

Install Software

Watch First Time Use Video
Onboarding Tutorials

Customize Role
Migrate Data
Add Employees

Read a Case Study
Text with Support
Watch or Listen to a Webinar
Ask a Question
Watch a Video
Read a FAQ
Talk to an Expert

Add on an App
Expand Seats
Expand Seats

User roles

Company owner
Bookkeeper - multiple companies, single company
Accountant - multiple companies
Payroll expert
Reporting expert
Controller
Business Analyst

Content

Available Recommended

Case Studies
Features FAQs

How-to Video
Onboarding Tips
Getting Started Guide

Case Studies
Webinars
How-to Videos
Support FAQs
Community Q & A
Tips
Newsletters

Expert Recommendations
Text Messages

User Goals

Get up to speed quickly

Enter Data Once

Manage Business Efficiently

First 90 days of a customer's journey with a small business product. What is shown doesn't capture the customer's emotional experience—that was on a different part of this diagram. Map by erin malone.

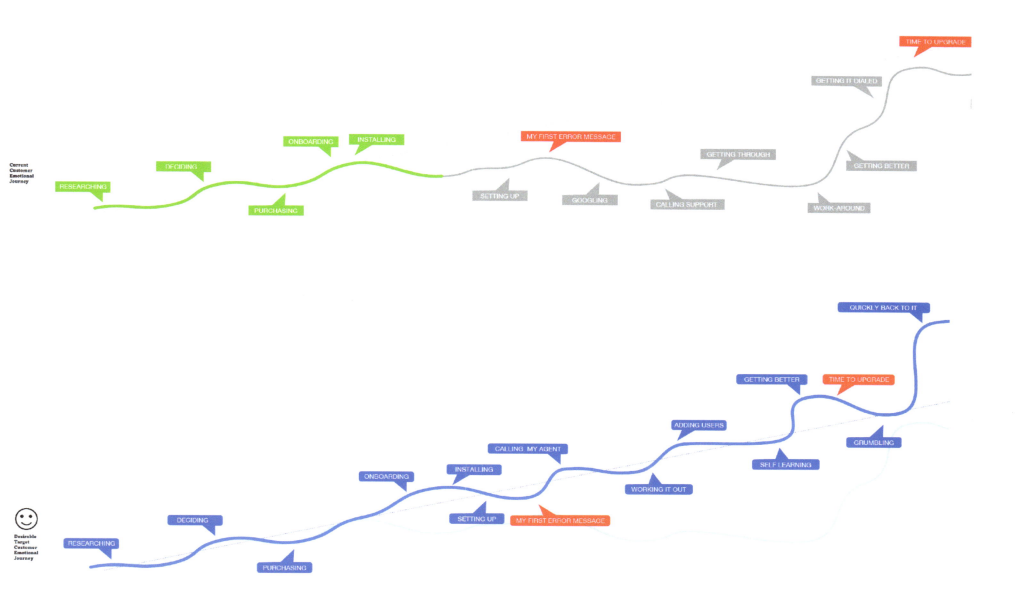

Current Customer Emotional Journey

RESEARCHING • DECIDING • ONBOARDING • INSTALLING • PURCHASING • SETTING UP • MY FIRST ERROR MESSAGE • GOOGLING • CALLING SUPPORT • GETTING THROUGH • WORK-AROUND • GETTING BETTER • GETTING IT DIALED • TIME TO UPGRADE

Desirable Target Customer Emotional Journey

RESEARCHING • DECIDING • PURCHASING • ONBOARDING • INSTALLING • CALLING MY AGENT • SETTING UP • MY FIRST ERROR MESSAGE • WORKING IT OUT • ADDING USERS • SELF LEARNING • GETTING BETTER • TIME TO UPGRADE • GRUMBLING • QUICKLY BACK TO IT

User hopes, goals and emotions. Top diagram documents the state of emotions before a major redesign to the site and the services offered. The bottom diagram captures the desirable emotional journey.

SERVICE BLUEPRINT

WHAT IS IT?

The service map is a map of the user journey across every touch point and channel and all the backstage—company & stakeholder activities—that happen at each point in the customer journey. The service blueprint helps teams coordinate processes and activities over time and allows teams to structure, design and align the business with the customer experience as well as giving a zoomed out view of the ecosystem.

WHEN DO YOU DO ONE?

The service design blueprint can begin during and completed after ethnography—both customer interviews and stakeholder interviews—and the team has a good understanding of current experiences, processes for the user as well as activities and business processes for the company.

HOW TO MAKE A SERVICE BLUEPRINT

This is going to be high-level. I recommend checking out the books This is Service Design Doing and This is Service Design Methods from O'Reilly Media for more in depth descriptions of service design research tools.

Make note of each step a user needs to take to complete their journey or experience with a service.

Make a list of what elements from the system, application or service they will interact with at each step.

Then make a list of every employee, stakeholder or behind the scenes person or tool that needs to be involved at each of these stages—this may include employees, technical systems and out of company stakeholders.

Draw a stack of swim lanes. In each lane along the left hand side, indicate the department or part of the service - both behind the scenes and in the user interaction that needs to be represented.

Then track the user's journey of interactions across the top. At each intersection, indicate what tasks or interactions need to happen to keep the user on task, meeting their goals or working through the experience.

On the next page: A portion of a service blueprint from a medical healthcare application on the next page shows the patient journey along the top and indicates each touch point and interaction in the application.

Below that swim lane are all the behind the scenes actors and systems involved that have to be activated in order to satisfy the user's goals. This includes doctors, nurses, pharmacists, customer support, insurance and out of system actors.

The diagram captures what was currently happening based on what was learned in the user research. Additionally, this document served as a guide to capture points of friction and inefficiencies and was used to indicate areas for improvement in the multiple interlocked systems. Map by erin malone.

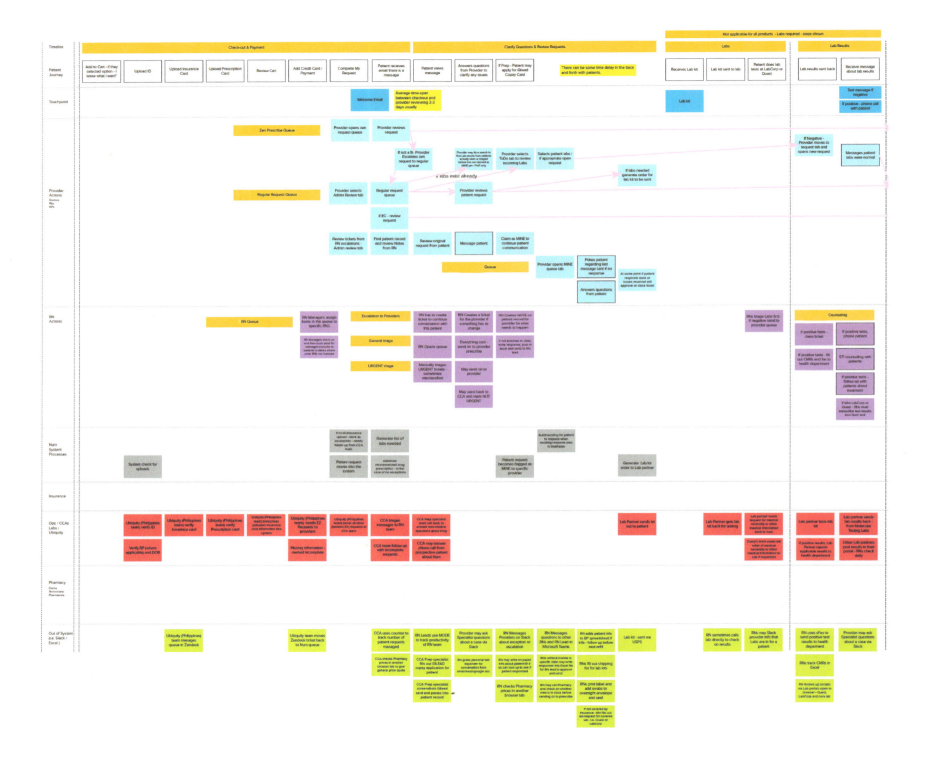

Customer lifecycle

Online discovery
Referral from (person)
Referral from (site post)
Referral from (ad)

Onboarding - mobile app

Lots of connecting and surfacing deals based on history and location

Only mobile?
Automate as much as possible - work less to setup

Personalize as much as possible - don't show stores not in geo

Mobile app
Pre shopping planning

Auto-create list from past history
Pre-populate with deals

Concierge service for convenience/less work based on system knowledge?

Mobile app
In store shopping

Alerts and list matching
Best deals
Brand switching opportunities
Upsell opportunities
Earn more - Pay with

Don't be creepy

Mobile app
Post shopping

Accounting/savings from purchases

Later purchase history & overall savings to date
Favorite brands

Map to online shopping if relevant i.e. shopped at Walmart, promote walmart.com & buy online
Prompt/triggers for next shopping trip

Post shopping
Prompt to refer to friends

Online use promotion

Remind about other online retailers - related items, people like you, brands also at...

Advertising upsell
Online upsell in google ads, fb and other social media ads

Online use

Search - metadata
Search results -
Listing
Page
Reviews
Most popular
Best deal
Brands

Post shopping
Prompt to refer to friends

Seasoned engaged user in store and online

Savings/earnings reports

Upcoming relevant deals

Layered deals (buy this too)

Partner engagement
Ad Banners
Video engagements (6m)
Offers
Recipes

Earn more - pay with ibotta

Post shopping
Prompt to refer to friends

Feedback loop
Trigger, action, reward

Feedback loop
Trigger, action, reward

Feedback loop
Trigger, action, reward

Feedback loop
Trigger, action, reward

Feedback loop
Trigger, action, reward

System stack

Activity stream | Personal feed

Contextual | Personalized shopping list

Search | Search Results

| Brands | Retailer | Savings | Pay with | additional savings | Popularity | Reviews | Price |

Notifications - triggers for action

| Brands specific | Location specific | Time specific | Savings specific | Store/Retailer specific |

Location

| Geo - nearby stores | Geo - nearby deals | Geo fence - where am I now? |

Personalization

| Bank info | Location info | Spending history | Brand preferences (inferred) | Profile |

Partner data

| Online Retailer Product Data | Exclusive content | Convert to walmart.com from shopping list add | Brand Deals (CPG) Data | Brand Engagement |

Customer lifecycle map over actions and activities that could happen at each stage. While not quite a full stack service blueprint, the goal was to understand the customer lifecycle and flesh out areas where the system and company could up sell, recommend partner data or content and other contextual and personalized benefits. Map by erin malone.

INTERACTION FLOWS

WHAT IS IT?

The interaction flow diagram breaks a large, complex application, site or service down into discrete parts allowing the designer to fully flesh out a single use case or user flow. The diagram captures decision points, alternate paths, system error triggers and can even capture interactions with the system database. An interaction flow may represent a piece of functionality contained within a "page" or may be a sequence of "pages" within a larger system.

A complex system project will have dozens of interaction flows diagramming each individual task a person might do in the system. These are also sometimes called User Flows or Task Flows.

WHEN DO YOU DO ONE?

These models are done once the team has an understanding of the users and their goals. These diagrams help flesh out functional ideas and expose issues. They are used in conjunction with product manager's requirements and engineering documents.

The interaction flow diagram helps facilitate the conversation of what to build and what needs to happen to create a great user experience as well as to document all aspects of the system. Think of it as a visualization of the conversation between the user and the system.

HOW TO MAKE AN INTERACTION FLOW

When I make interaction flows I start out by listing each possible task a
person might need to do in the system.

These might be things like:
Signing Up
Signing In
Getting a Lost Password
Getting a Lost User Name
Uploading a Document or Image
Drawing on something
Saving something
Sharing something
etc.

The list goes on depending on your project and system.

A flow can also be done to represent actions of stakeholders in the system
if they are part of the user's flow.

Then I list out each step for each task. For example, something simple like sign in:
Go to the page (or module)
Enter user name
Enter password
Select submit

Now the system needs to respond back.
Did the person put in the right user name?
The right password?
The right format?

If everything is fine, the happy path as some call it, the user goes on their way.

If everything is not fine, then the system sends the user back and shows appropriate error messages to help them correct the issues. For some things like Sign Up or Sign In, client side checks may happen as a person tabs out of a field (i.e. format may not be correct) and the error messages show up before submitting the form, alleviating the need for a server call.

When I draw out an interaction flow, the boxes represent the stage of the task, and diamonds represent decision points. Points where the SYSTEM—not the user—has to decide if the information is correct or not and make a decision, show an error or let them go on their way.

Decision points are usually addressed with a question that can be answered with YES or NO.

If NO there is usually some other path or a return to the starting point. If YES the task proceeds to the next step.

Figuring out what a system decision entails is where conversations with your engineering partners is helpful. Is there a database or server call involved here? Is it just a hyperlink going to the next html page? These should be represented differently.

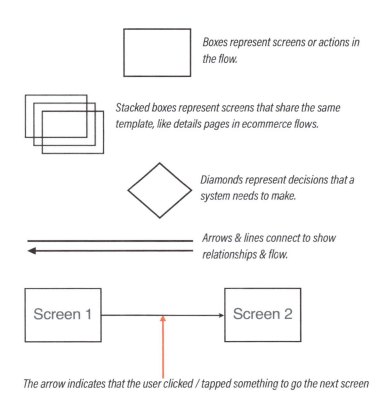

Boxes represent screens or actions in the flow.

Stacked boxes represent screens that share the same template, like details pages in ecommerce flows.

Diamonds represent decisions that a system needs to make.

Arrows & lines connect to show relationships & flow.

Screen 1 → Screen 2

The arrow indicates that the user clicked / tapped something to go the next screen

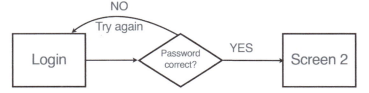

The diamond indicates a system decision.
There should be a YES path - successful outcome of the question
- and a NO path if the decision result fails.

New User Sign Up - Flow - Email as UserID

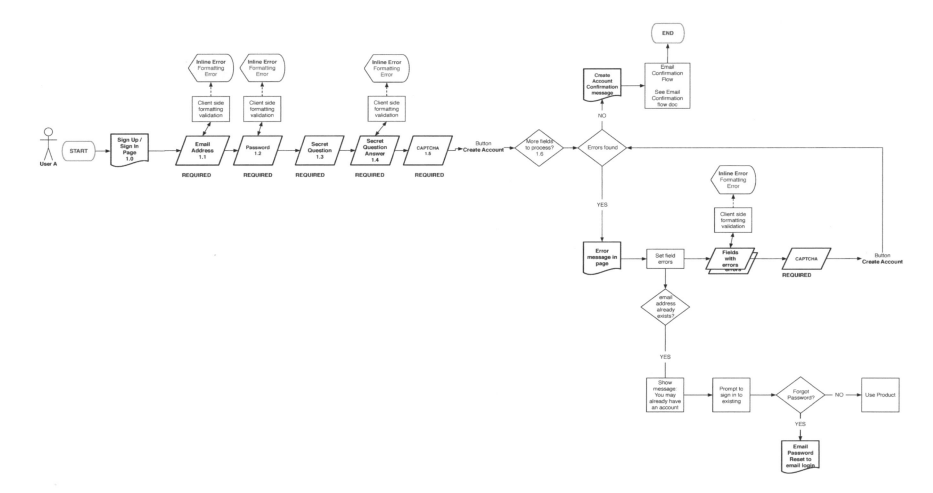

Interaction flow for signing in to a site. In this case I used angled boxes to show each part of the task and plain boxes to indicate client side checks or server side checks. The rounded arrow boxes indicate what type of error message to show.

As you can see, something as simple as signing into a site, is actually not as simple as you think. Interaction flow diagram by erin malone..

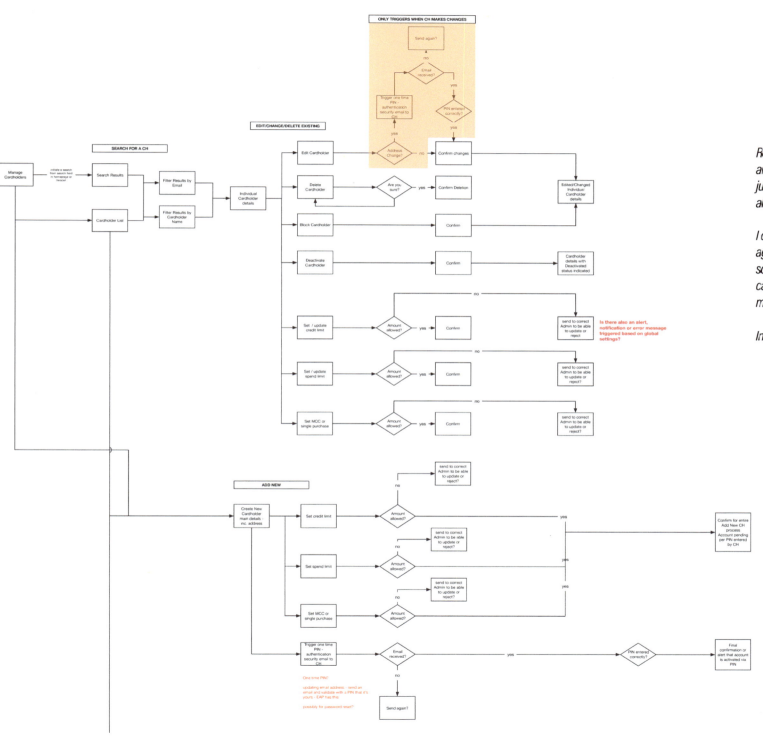

Robust interaction flow for a major section of an enterprise banking application. This was just one page of 8 interaction flows covering all the functionality in the product.

I often leave notes in red for product management and development as I am working so that we can discuss if the interaction can actually be built or if there are elements missing.

Interaction flow diagram by erin malone.

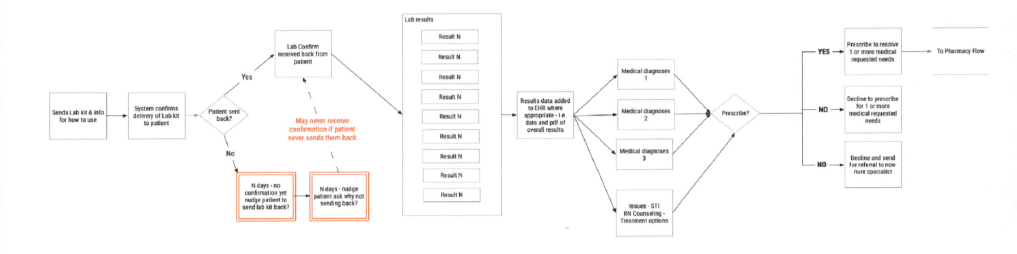

Sends Lab kit & info for how to use → System confirms delivery of Lab kit to patient → Patient sent back?

Patient sent back? — Yes → Lab Confirm received back from patient

Patient sent back? — No → N days - no confirmation yet nudge patient to send lab kit back? → N days - nudge patient ask why not sending back?

May never receive confirmation if patient never sends them back

Lab results
- Result N
- Result N
- Result N
- Result N
- Result N
- Result N
- Result N
- Result N
- Result N

→ Results data added to EHR where appropriate - i.e. data and pdf of overall results →

- Medical diagnoses 1
- Medical diagnoses 2
- Medical diagnoses 3
- Issues - STI RN Counseling - Treatment options

→ Prescribe? →

YES → Prescribe to resolve 1 or more medical requested needs → To Pharmacy Flow

NO → Decline to prescribe for 1 or more medical requested needs

NO → Decline and send for referral to non-nurx specialist

One of many task flows for a healthcare provider. This one walks through the process of a person receiving a lab kit and then sending it back to the provider and then the process that is involved.

This particular flow involves people, real objects, multiple stakeholders as well as parts of the digital system. Task flow by erin malone.

INFORMATION MODEL

WHAT IS IT?

A model to define the information, content or data in a system and to indicate relationships between each entity. Sometimes called a Content Model (if done by a Content Strategist). The information model can also be called a Data Model and documents data entities, attributes and their relationships. The data model - content model - information model may be co-developed and refined across teams. The information model may appear in the form of both a diagram depicting relationships and a spreadsheet defining all the relevant entities and attributes.

WHEN DO YOU DO ONE?

Information models are started at the beginning of the information architecture and content strategy phase. Early versions can model the containers and relationships while later versions are refined and might contain more specific content and metadata.

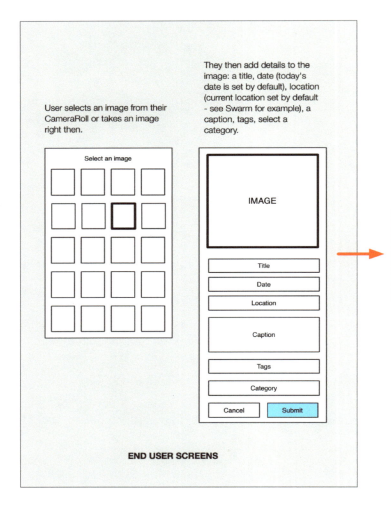

User selects an image from their CameraRoll or takes an image right then.

They then add details to the image: a title, date (today's date is set by default), location (current location set by default - see Swarm for example), a caption, tags, select a category.

Select an image

IMAGE

Title

Date

Location

Caption

Tags

Category

Cancel   Submit

END USER SCREENS

Upon UPLOAD a unique image identifier number is associated with the image so that titles, user names and other info doesn't have to be unique. Additionally, UPLOAD DATE is added to the metadata

If the image is published, a PUBLISH DATE is added to the database so that the image might be filtered out of the pool for future selections.

Image Data
Title: string
Image_Date: Date
Location: string, GPS coordinates
Caption: string
Tags: comma delimited string
Category: string
Image: jpg or png
User_Name: string

Image Data
Unique_Image_Identifier: Number
Title: string
Image_Date: Date
Location: string, GPS coordinates
Caption: string
Tags: comma delimited string
Category: string
Image: jpg or png
User_Name: string
Upload_Date: Date

Image Data
Unique_Image_Identifier: Number
Title: string
Image_Date: Date
Location: string, GPS coordinates
Caption: string
Tags: comma delimited string
Category: string
Image: jpg or png
User_Name:string
Upload_Date: Date
Publish_Date: Date

METADATA SPECS

Data model for a conceptual application. The metadata in the charts maps to the data collected by the user interface AND the inherent metadata attached to the EXIF data in the image uploaded. This information will be used for filtering and sorting as well as presenting dynamic content in other parts of the application. The presentation shows the database labels as well as the specs for the data to be collected. This will help an engineer create the right database structure. This content shown as well as other content needed throughout the application may also be tracked in a spreadsheet.
Model by erin malone.

USE CASE DIAGRAMS

WHAT IS IT?

A use case diagram presents the tasks a person wants or needs to do and identifies the relationships between the tasks and the user. These sometimes look similar to concept models but can be linear or hierarchical depending on the tasks and how granular they become in the documentation. Just because something appears in a use case diagram, doesn't mean it will or should be supported in an experience or application.

Working through the diagram also helps define the boundaries of the system and all the actors who need to be considered in the system.

You may see these also referred to as UML (Unified Modeling Language) Use Case diagrams.

WHEN DO YOU DO ONE?

Early in the process to help understand the users and their needs as related to a set of tasks you are looking to support. There might be separate diagrams for each user type or large suite of tasks.

HOW TO MAKE A USE CASE MODEL

Write out the goals and steps of the task your customers need to accomplish. This might come out like a written scenario or narrative or could be a list of tasks by category.

Then plot out each step that needs to happen to accomplish that task. This can include steps that in the software or digital product as well as steps that might need to happen outside the digital experience. While this isn't an experience map, its helpful to understand and capture context.

Next, define each actor in the system. This might be a first time user or a repeat user or a support agent or all three. We use stick figures to represent the actors. These are nouns in the system.

Start with the stick figure and move to the right with each step of each task. Connect the actor to the steps with the verbs related to that relationship (see how this is similar to concept modeling). The actors may be connected together by the flow - i.e. author to editor with tasks between connecting them.

Key Tasks

Create / Save

Account
Campaign
Ad Group
Ad
Keywords
Bids
Reports

Edit

Campaign
Ad Group
Ad
Keywords
Bids
Schedule
Reports
Account

Review

Reports
Campaigns
Ad Groups
Ads
Keywords
Bids
Schedule
Position

A list of tasks and parent tasks (verbs) related to the system for bidding on ads.

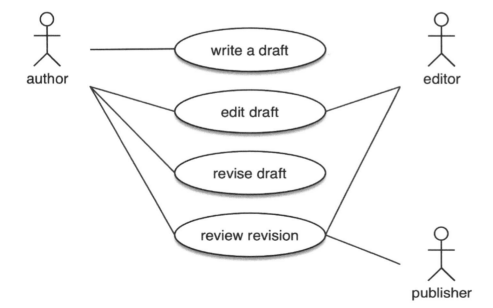

A super simplified use case for the tasks and actors in the system of writing an article. The actors are the author, the editor and the publisher. The tasks are writing, editing, revising, reviewing and eventually publishing (not shown).

First Time User

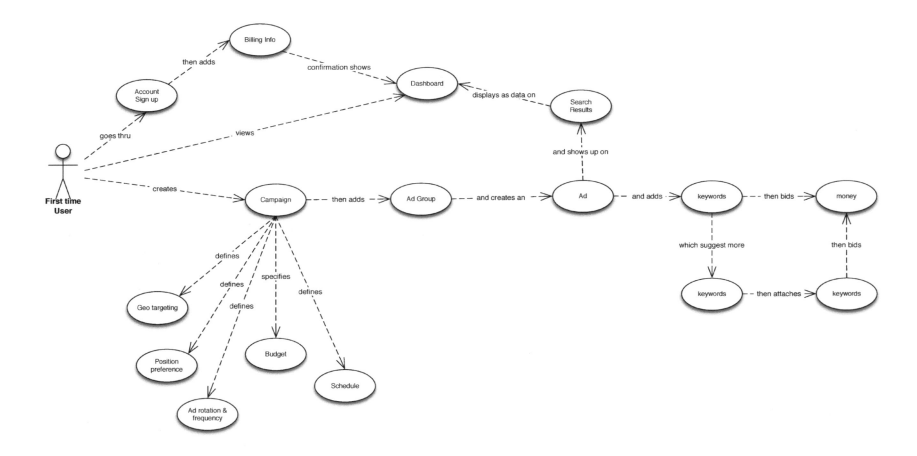

Use case diagram for a first time user of an ad bidding system.
Diagram by erin malone.

Return User - Small Customer

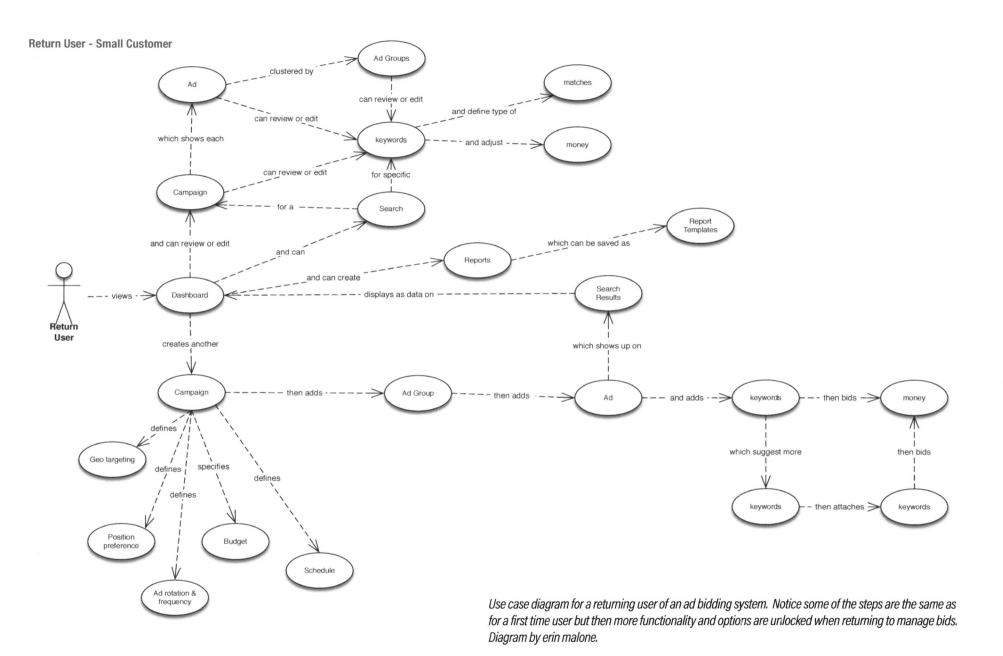

Use case diagram for a returning user of an ad bidding system. Notice some of the steps are the same as for a first time user but then more functionality and options are unlocked when returning to manage bids. Diagram by erin malone.

**Search
by keyword**

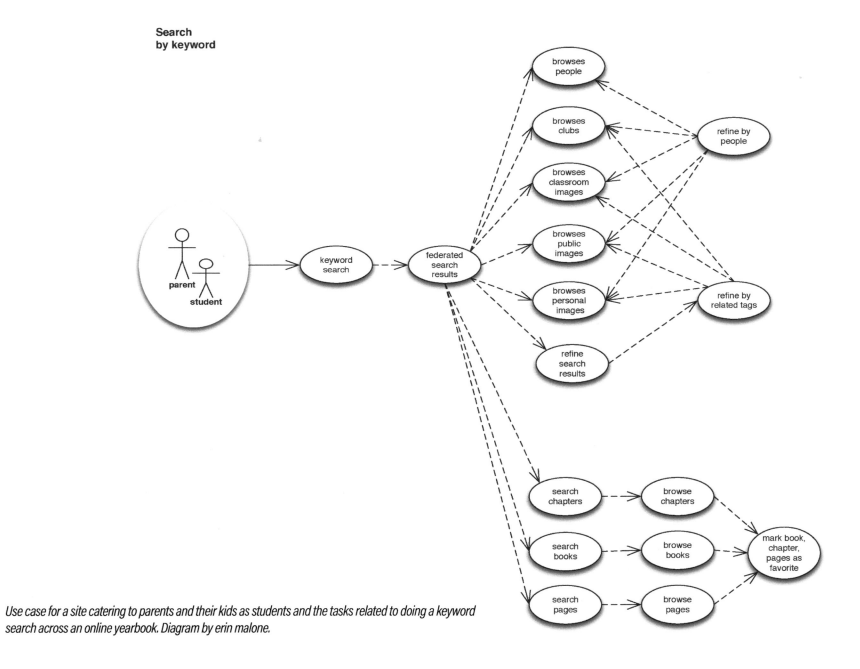

*Use case for a site catering to parents and their kids as students and the tasks related to doing a keyword
search across an online yearbook. Diagram by erin malone.*

Acquiring Images

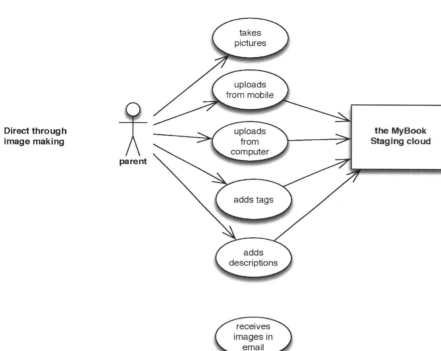

Direct through image making

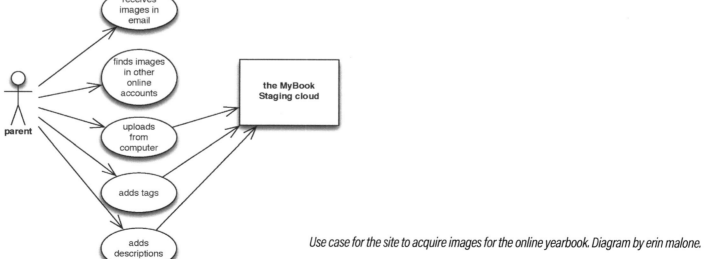

Indirect through email and other websites (still outside the system)

Use case for the site to acquire images for the online yearbook. Diagram by erin malone.

TIMELINE DIAGRAMS

WHAT IS IT?

A Timeline diagram shows activities and actions across time. These may be done as a final deliverable or may be done to track a project in progress. Gantt Charts are a type of timeline diagram generally used by project managers to track a project over a set period of time with milestones and deliverables called out. A timeline may also be done to show project process and key milestone dates. It may also be done to explore the work and life of a person, company or other organization as a final artifact in its own right as opposed to being used as a communication tool with team stakeholders.

WHEN DO YOU DO ONE?

For project management and project tracking, a timeline may be done before a project starts to help capture projected milestones and deliverables to achieve team buy in on key dates. They may also be done as part of a project estimation or proposal process.

External timelines may be done as part of a website or print deliverable and generally is populated with content from an editorial or content team.

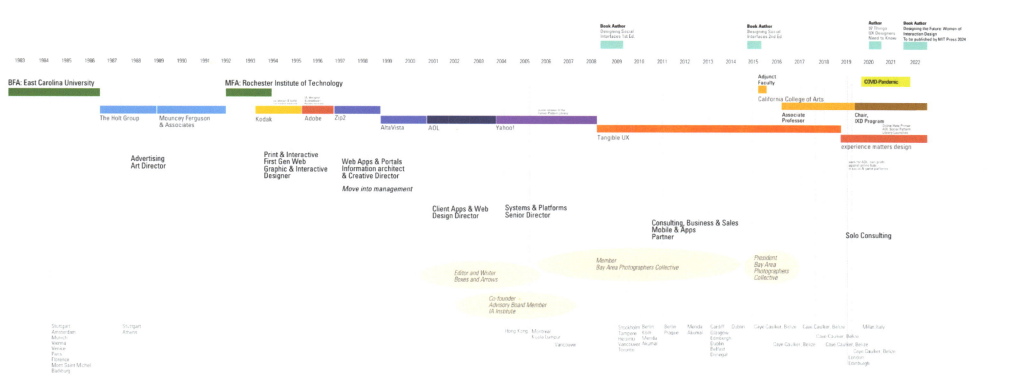

A timeline diagramming my professional career. The timeline documents my education, work entities and types of work done, professional affiliations and teaching engagements. For fun I also indicated international trips and my book writing and publishing efforts. I have published this on my website and often start off presentations with this diagram when introducing myself although now its starting to show its age. Model by erin malone.

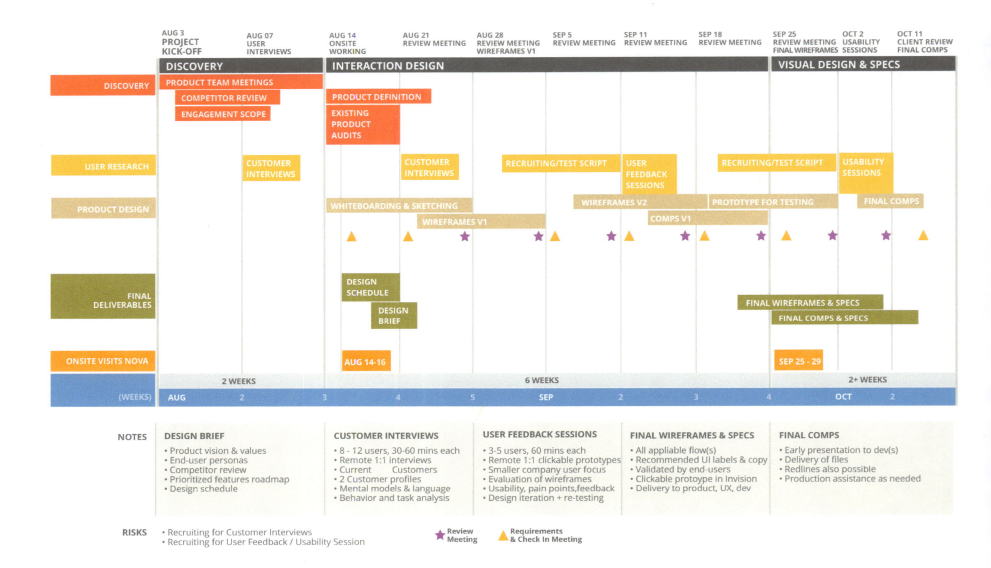

	AUG 3 PROJECT KICK-OFF	AUG 07 USER INTERVIEWS	AUG 14 ONSITE WORKING	AUG 21 REVIEW MEETING	AUG 28 REVIEW MEETING WIREFRAMES V1	SEP 5 REVIEW MEETING	SEP 11 REVIEW MEETING	SEP 18 REVIEW MEETING	SEP 25 REVIEW MEETING FINAL WIREFRAMES	OCT 2 USABILITY SESSIONS	OCT 11 CLIENT REVIEW FINAL COMPS

DISCOVERY — **INTERACTION DESIGN** — **VISUAL DESIGN & SPECS**

DISCOVERY
- PRODUCT TEAM MEETINGS
- COMPETITOR REVIEW
- ENGAGEMENT SCOPE
- PRODUCT DEFINITION
- EXISTING PRODUCT AUDITS

USER RESEARCH
- CUSTOMER INTERVIEWS
- CUSTOMER INTERVIEWS
- RECRUITING/TEST SCRIPT
- USER FEEDBACK SESSIONS
- RECRUITING/TEST SCRIPT
- USABILITY SESSIONS

PRODUCT DESIGN
- WHITEBOARDING & SKETCHING
- WIREFRAMES V1
- WIREFRAMES V2
- PROTOTYPE FOR TESTING
- FINAL COMPS
- COMPS V1

FINAL DELIVERABLES
- DESIGN SCHEDULE
- DESIGN BRIEF
- FINAL WIREFRAMES & SPECS
- FINAL COMPS & SPECS

ONSITE VISITS NOVA
- AUG 14-16
- SEP 25 - 29

2 WEEKS — 6 WEEKS — 2+ WEEKS

(WEEKS)	AUG	2	3	4	5	SEP	2	3	4	OCT	2

NOTES

DESIGN BRIEF
- Product vision & values
- End-user personas
- Competitor review
- Prioritized features roadmap
- Design schedule

CUSTOMER INTERVIEWS
- 8 - 12 users, 30-60 mins each
- Remote 1:1 interviews
- Current   Customers
- 2 Customer profiles
- Mental models & language
- Behavior and task analysis

USER FEEDBACK SESSIONS
- 3-5 users, 60 mins each
- Remote 1:1 clickable prototypes
- Smaller company user focus
- Evaluation of wireframes
- Usability, pain points, feedback
- Design iteration + re-testing

FINAL WIREFRAMES & SPECS
- All appliable flow(s)
- Recommended UI labels & copy
- Validated by end-users
- Clickable protoype in Invision
- Delivery to product, UX, dev

FINAL COMPS
- Early presentation to dev(s)
- Delivery of files
- Redlines also possible
- Production assistance as needed

RISKS
- Recruiting for Customer Interviews
- Recruiting for User Feedback / Usability Session

★ Review Meeting   ▲ Requirements & Check In Meeting

Example project schedule created once the project was secured by the design firm. The schedule shows information from the UX design perspective and only includes UX design initiatives by role, each in their own swim lane, deliverable milestones along with details of what each activity entails.

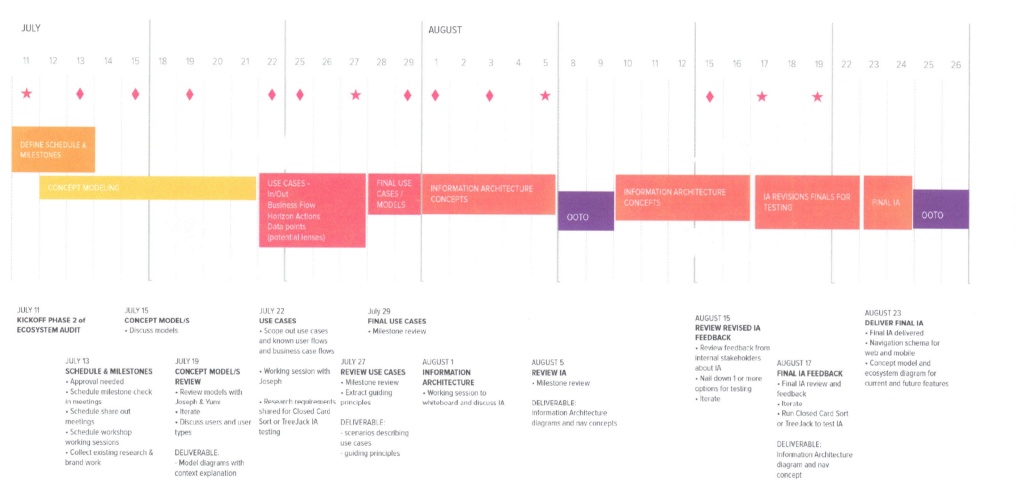

JULY **AUGUST**

| 11 | 12 | 13 | 14 | 15 | 18 | 19 | 20 | 21 | 22 | 25 | 26 | 27 | 28 | 29 | 1 | 2 | 3 | 4 | 5 | 8 | 9 | 10 | 11 | 12 | 15 | 16 | 17 | 18 | 19 | 22 | 23 | 24 | 25 | 26 |

DEFINE SCHEDULE & MILESTONES

CONCEPT MODELING

USE CASES - In/Out Business Flow Horizon Actions Data points (potential lenses)

FINAL USE CASES / MODELS

INFORMATION ARCHITECTURE CONCEPTS

OOTO

INFORMATION ARCHITECTURE CONCEPTS

IA REVISIONS FINALS FOR TESTING

FINAL IA

OOTO

JULY 11
KICKOFF PHASE 2 of ECOSYSTEM AUDIT

JULY 13
SCHEDULE & MILESTONES
• Approval needed
• Schedule milestone check in meetings
• Schedule share out meetings
• Schedule workshop working sessions
• Collect existing research & brand work

JULY 15
CONCEPT MODEL/S
• Discuss models

JULY 19
CONCEPT MODEL/S REVIEW
• Review models with Joseph & Yumi
• Iterate
• Discuss users and user types

DELIVERABLE.
- Model diagrams with context explanation

JULY 22
USE CASES
• Scope out use cases and known user flows and business case flows

• Working session with Joseph

• Research requirements shared for Closed Card Sort or TreeJack IA testing

JULY 27
REVIEW USE CASES
• Milestone review
• Extract guiding principles

DELIVERABLE:
- scenarios describing use cases
- guiding principles

July 29
FINAL USE CASES
• Milestone review

AUGUST 1
INFORMATION ARCHITECTURE
• Working session to whiteboard and discuss IA

AUGUST 5
REVIEW IA
• Milestone review

DELIVERABLE:
Information Architecture diagrams and nav concepts

AUGUST 15
REVIEW REVISED IA FEEDBACK
• Review feedback from internal stakeholders about IA
• Nail down 1 or more options for testing
• Iterate

AUGUST 17
FINAL IA FEEDBACK
• Final IA review and feedback
• Iterate
• Run Closed Card Sort or TreeJack to test IA

DELIVERABLE:
Information Architecture diagram and nav concept

AUGUST 23
DELIVER FINAL IA
• Final IA delivered
• Navigation schema for web and mobile
• Concept model and ecosystem diagram for current and future features

◆ Working Sessions: Brainstorms & collaborative sessions; Workshops   ★ Review meeting: Walk through ideas, Collect feedback, next steps

Example project schedule created once the project was secured by the design firm. The schedule shows information from the UX design perspective and only includes UX design initiatives and milestones as well as details of what each activity entails.

With this guide, I have tried to showcase a variety of maps, models and diagrams. I have included the ones I use most in my work as a user experience designer. There are others that I have not included, including Empathy Maps and Mental Models. Mostly because I have never actually done them in my work, and I wanted to be able to illustrate the guide with my own examples.

For many of these kinds of maps you can find blank templates created in Mural and Miro and other tools. All you need to do is fill in the boxes or ovals. In almost every case, though, the model or map should be preceded by robust and thorough ethnography including stakeholder interviews, audits of existing products and services and competitive analysis.

These models are intended to aid in your understanding of the systems you are to be designing within, to give you context and a scope of the boundaries involved. They range from conceptual and exploratory to descriptive and evaluative. Different models help you at different phases. Some are just for you, to get your head around a space, and others are intended for communication with other team members.

MORE RESOURCES

Aronson, Daniel. "Step-By-Step Stocks and Flows: Improving the Rigor of Your Thinking." The Systems Thinker, January 12, 2016. http://thesystems-thinker.com/step-by-step-stocks-and-flows-improving-the-rigor-of-your-thinking/.

Brown, Daniel M. *Communicating Design: Developing Web Site Documentation for Design and Planning*. Berkeley, Ca: New Riders, 2011.

Buzan, Tony. *The Ultimate Book of Mind Maps*. Harper Collins UK, 2012.

Covert, Abby. *Stuck? Diagrams Help*. Abby Covert, 2022.

Covert, Abby. *How to Make Sense of Any Mess*. Abby Covert, 2014.

Duarte, Nancy. slide:ology: *The Art and Science of Presentation Design*. Sebastopol: O'Reilly Media, 2011.

Dubberly, Hugh. "Creating Concept Maps." Dubberly Design Office. Accessed April 16, 2023. http://www.dubberly.com/concept-maps/creating-concept-maps.html.

Faste, Rolf. "Mind Mapping," 2011. http://www.fastefoundation.org/publications/mind—mapping.pdf.

Kalbach, James. *Mapping Experiences*. Sebastopol: O'Reilly Media, 2020.

Meadows, Donella H. *Thinking in Systems: A Primer*. Illustrated edition. White River Junction, Vermont: Chelsea Green Publishing, 2008.

Morville, Peter. *Ambient Findability*. Sebastopol: O'Reilly Media, 2005.

Stickdorn, Marc. *This Is Service Design Doing, Applying Service Design Thinking in the Real World: A Practitioners' Handbook*. Sebastopol: O'Reilly Media, 2018.

Stickdorn, Marc, Adam Lawrence, Markus Hormess, and Jakob Schneider. *This Is Service Design Methods, a Companion to This Is Service Design Doing : Expanded Service Design Thinking Methods for Real Projects*. Sebastopol: O'Reilly Media, 2018.

Stroh, David Peter. *Systems Thinking for Social Change : A Practical Guide to Solving Complex Problems, Avoiding Unintended Consequences, and Achieving Lasting Results*. White River Junction, Vermont: Chelsea Green Publishing, 2015.

The Systems Thinker. "Systems Thinking Tools: A User's Reference Guide," March 9, 2016. https://thesystemsthinker.com/systems-thinking-tools-a-users-reference-guide/.

Unger, Russ, and Carolyn Chandler. *A Project Guide to UX Design : For User Experience Designers in the Field or in the Making*. Berkeley, Ca.: New Riders, 2012.

Wodtke, Christina. "How to Make a Concept Model". Boxes and Arrows, May 6, 2014. http://boxesandarrows.com/how-to-make-a-concept-model/

Wodtke, Christina. "Five Models for Making Sense of Complex Systems". Medium, February 11, 2017. https://cwodtke.medium.com/five-models-for-making-sense-of-complex-systems-134be897b6b3#.84emzqh4t

Wodtke, Christina. "A Visual Vocabulary for Concept Models". Elegant Hack, May 12, 2017. https://eleganthack.com/a-visual-vocabulary-for-concept-models/

Young, Indi. *Mental Models : Aligning Design Strategy with Human Behavior*. Brooklyn, N.Y. Rosenfeld Media, 2008.

Zak, Josh. The Ultimate Guide to Experience Mapping. https://www.experiencemap.com/

ABOUT ERIN MALONE

Erin Malone has over 25 years of experience leading experience design teams, deconstructing complex systems, designing web and software applications, and social experiences.

She is a Professor and the Chair of the Interaction Design BFA program at California College of the Arts. She teaches IXD Foundations, Visual Interaction Design, Systems Thinking, Interaction Design History and Behavior.

She was recently a principal UX Designer & Researcher at Experience Matters Design, a UX consulting firm specializing in designing systems, complex internal tools and social interfaces, as well as providing research and creating products that matter to under-serviced populations. Her most recent projects involved deconstructing the ecosystem of online hate in social platforms and games for the Anti Defamation League.

Erin holds an MFA in New Media and Graphic Design from Rochester Institute of Technology and a BFA in Graphic Design from East Carolina University's School of Art.

www.ingramcontent.com/pod-product-compliance
Lightning Source LLC
LaVergne TN
LVHW072101060326

832903LV00055B/379